Recurring Logistic Problems

As I Have Observed Them

GENERAL MAGRUDER

Recurring Logistic Problems

As I Have Observed Them

by

Carter B. Magruder
General, U.S. Army, Retired

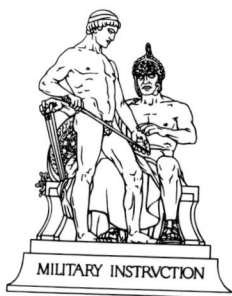

MILITARY INSTRVCTION

Center of Military History
United States Army
Washington, D.C., 1991

Library of Congress Cataloging in Publication Data

Magruder, Carter B. (Carter Bowie)
    Recurring logistic problems as I have observed them / by Carter B.
Magruder.
      p. cm.
    1. United States. Army —Supplies and stores—History—20th
century. 2. United States. Army—Transportation—History—20th
century. 3. World War, 1939–1945—Logistics—United States.
4. Magruder, Carter B. (Carter Bowie) I. Title.
UC23 1941–1969.M34 1990
355.4'11'0973—dc20

90–1928
CIP

CMH Pub 70-39

First Printing

For sale by the Superintendent of Documents, U.S. Government Printing Office
Washington, D.C. 20402

Foreword

The study of the logistical aspects of war is of particular importance in our peacetime Army because, as General Carter B. Magruder so forcibly reminds us in the pages that follow, basic problems tend to recur in logistics. Despite the radical transformation in equipment and supplies that separate today's Army from that entered by Magruder in 1917, the principles that guided the technical services of his day apply equally to those who serve in combat service support assignments in 1989. But if the principles have endured, so too, as Magruder demonstrates, have the problems. It was General Magruder's hope, as it is of those historians who study this vital segment of the military craft today, that by examining these problems within the context of military operations, improvements can be achieved.

I am pleased to be able to publish this important historical essay. Long buried in the Department of the Army's archives and known but to a handful of researchers, it will now be available to military students as well as to scholars who are increasingly coming to understand that the study of the logistical aspects of war is vital to an understanding of our military past.

Washington, D.C.

HAROLD W. NELSON
Colonel (P), USA
Chief of Military History

Preface

My purpose in this prefatory note is to introduce the author, explain why and how the book was written, and in a general way provide signposts to guide the reader on his journey. General Carter Bowie Magruder, United States Army, was one of the last survivors of the cadre of planners and commanders who trained, deployed, supplied, and committed in global battle the millions of American troops raised by this nation in World War II. When, in 1961, Magruder retired from his last post (Commanding General, U.S. Forces in Korea and Eighth U.S. Army) he was recognized as the top logistician in the Army. During the final twenty years of his long military career he had acquired an intimate and accurate knowledge of the extraordinary tasks involved in providing support to troops fighting in theaters of war thousands of miles apart and a great distance from the American home bases. The span of Magruder's experience and knowledge extended from the preparations for the invasion of Northwest Africa (TORCH) to those last days when American troops were departing Vietnam. The illustrations he draws come from very diverse sources: witness the provision of support for a field artillery gun team and the contrasting employment of computer simulation to evaluate "worthwhile targets" and determine ammunition requirements to engage the same.

After General Magruder took his retirement parade, the many years of his uniformed career given to logistics were augmented by a new role as a consultant for the Research Analysis Corporation (RAC). The apparition of the atomic bomb had forced the Army to recruit civilian physicists and mathematicians who could explain this new weapon and its effects. The Johns Hopkins University created a not-for-profit "think tank" at the Army's request to house these academicians and theorists, the Operations Research Office (ORO). Over the years the analytical methods applied to atomic studies became useful in other areas of Army concern, particularly in the fields of computer simulation, war-gaming, and logistics. When ORO was expanded to become the Research Analysis Corporation—still a not-for-profit organization dedicated to Army problems—a considerable group of prestigious retired officers joined RAC, attracted by the opportunity to work on real and pressing Army problems. Magruder was part of this group; he had not lost his interest in the perennial problems inherent in the Army's logistical structure when he took off his uniform. The general now was a part of a research team, with no staff at his command, but he did have, and relished the opportunity to give free play to his ideas and to exploit the "possibilities" afforded by the high-speed digital computer. He enjoyed the intellectual challenge of these years, as this book clearly shows. But in the process of time, Magruder was ready for a final retirement, to play golf, join his old partners at bridge, and spend time with his adored wife.

Now I must interject myself into this introduction (requesting the reader's indulgence) and explain how Carter's book came to be. His competence and expertise in the most complex areas of logistics seemed to dictate that this experience should be preserved for a new generation of Army logisticians, both uniformed and civilian. A written account, prepared by the general at the close of his career, could, it seemed to me, be modeled on the Final Report that is written when a theater of operations is closed and which is designed to summarize problems encountered and lessons learned. Such a piece, however, was not covered by the Army's contract with RAC, but I had a few "discretionary" dollars.

Magruder was the soul of precision, so he received a formal memorandum which requested that he write a "think piece" that would "plumb all of your experience and your past thinking about Army logistics and distill . . . your personal findings and conclusions." As a selling point, I reminded the general of the "simple fact of life that the Army high-level decision-makers and planners do not have time and opportunity to do the kind of thing that you might be able to do." When he accepted, I formed a review board from the top operations analysts in RAC, all with wartime experience, and from his pcers among our consultants: retired Army Generals Thomas T. Handy, James E. Moore, and Charles D. Palmer. These latter three and Magruder normally acted in concert to give RAC programs in war-gaming and computer simulations a military depth and breadth that would have been hard to equal, then and now.

Early in 1969, Magruder completed his first draft chapters. But there was a problem. The general had a reputation as a severe and rigorous taskmaster (those who witnessed his dealings with service chiefs when they became obstructionists can attest that Carter was not a pliable *persona*). Magruder, however, was a very courtly gentleman of the old school who regarded any hint of personal aggrandizement as unbecoming an officer in the United States Army. Not surprisingly, his chapters were written in the third person and almost totally devoid of his own thumb prints. A second memorandum to Magruder contained a more precise statement of the "Object of the Exercise" (to use the old Leavenworth phrasing). He was *not* asked to write a field manual, nor was he to seek to encompass the entire field of logistics. Completeness, I suggested, was not a virtue except in textbooks. Very simply, he was being asked to say in print: "This is what I, Carter Magruder, now believe to be important in the area of logistics." Fortunately, General Handy, recognized as first among equals, was head of the special review board named for Magruder's manuscript. Handy took up the cudgels to get what he called "the most *Magruder*" into each successive revision. In time, Carter became less inimical to the words "I" and "my."

There remained one last hurdle. The title given the completed book, in manuscript, was suggested by General Handy: "Recurring Logistic Problems As I Have Observed Them." This, at bottom, was what the book was about and seemed a clear statement that Carter spoke for himself alone. But at this time the war in Indochina had reached a peak, as had demonstrations against it back in the United States. Magruder became gravely concerned that the critics of Americans in uniform might somehow find a handhold in his book, or that he might offend serving officers who bore the heat of the day. I asked the two other members of the "4 Star" quadrivium to carefully read the final manuscript with Magruder's fears in mind. Moore and Palmer agreed that the book posed no such problems. General Palmer, with his usual trenchant acerbity, observed that Carter's comments were worthwhile because ". . . they might cause some justifiable and possibly useful soul-searching in the Army."

Magruder's task was accomplished. But RAC was in the throes of dissolution as the Army sought to bring operations research and analysis "in house." (It may be of interest to students of our national economy that I needed only $2,500 to publish the book in 1970, but no credit was extended.) The general's final manuscript disappeared into the Army's archival maw. Now, two decades later, Dr. Joel Meyerson, an Army historian preparing the official history of logistics in the Vietnam War, rediscovered Carter Magruder's last will and testiment to the Army he had served so faithfully. The Center of Military History turned to a selection of serving general officers assigned in logistics commands for a new critique of the Magruder work. (In retrospect, one may conclude that no book ever has been scrutinized by as many "stars" as this.) The book is being published as Magruder wrote it, with his inflections and conclusions. His format shows the logical mind for which he was famous. His organization of subject matter leads, in each case, to two specific summary statements: *reasons* why the (logistic) problem recurs, and the *form* in which the problem probably will recur.

The action officer working on a current contingency plan or the service school instructor trying to explain the wartime importance of the "repair parts problem" surely will find much in this book to illuminate and enlarge his understanding of his assigned task. But all may profit from General Carter Magruder's final words on integrity and the soldier; so said his peers twenty years ago.

October 1989   HUGH M. COLE

Author's Preface

> I have but one lamp by which my feet are guided,
> and that is the lamp of experience. I know of no
> way of judging the future but by the past.
>
> Patrick Henry

It is my objective in this paper to identify areas in which I have observed Army logistic problems recur, and then to review some of the specific problems that I have seen arise in each area, giving something of their history, the solutions attempted, the degree of success attained, and the reasons why I think they may arise again. This is done with the hope that some conceptions from my experience, however limited, may help others in the solution of future logistic problems.

Shortages beget logistics problems. Critical problems arise and recur therefore with greater frequency when a maximum effort is being made. Korea, like Vietnam, was a limited war, a "guns and butter" war. In neither was the economy strained, so problems were less critical and less frequent. World War I was a maximum effort, but my experience in it was brief and only as a reserve second lieutenant of infantry. As a result, most of my examples are taken from World War II. World War II was a conventional war except in its last phase. Since World War II much of our military concern has focused on nuclear weapons, with some concern for chemical and biological weapons. I do not feel overwhelmed at the thought of nuclear war. Bearing in mind the precept of that grand old man Tom Jenkins, who taught us wrestling when I was a cadet at West Point, that "there ain't no hold that can't be broken," I feel that countermeasures will be developed to moderate the effect of nuclear weapons. I also think that until this is accomplished no great nation will risk such a war. Accordingly, I feel that both peacetime logistics and logistics in future wars may well resemble logistics in early wars. The orientation of my thoughts on logistics is thus toward a major war in the future regardless of what weapons may be used.

I have devoted a chapter to each area I have identified as one in which logistic problems recur. Within each chapter, I have stated the reasons why I believe the problems in the logistic area being covered recur, together with some indication as to the form and manner in which, when they recur, they may be presented to Army logisticians. My final chapter has been named both a statement of lessons learned and a summary. As a matter of fact, it is not a good example of either. What I have really done is to set forth what I believe to be important in logistics. To the extent that most of my so-called lessons learned are to some degree supported in my discussions of various recurring problems, it may be considered something of a summary. To the extent that I have stated conceptions that are generally true—although put in a simple form without qualifications in order to secure emphasis—it may be considered something of a statement of lessons learned.

Having never intended to write a book or even a paper such as this one, I have kept no files from my active service. The accuracy of the statements made herein is, therefore, subject to the fallibility of my memory.

This paper is based primarily upon some twenty-eight years of work devoted, in major part, to logistics. It includes my last twenty years of active duty in the Army, starting with my assignment to the G-4 Division, War Department General Staff, in 1941, and ending with my retirement in 1961. It also includes the succeeding eight years spent as a consultant on logistics to the Research Analysis Corporation and to the Logistics Management Institute. In order to indicate what opportunities I have had to see problems recur in logistics and to develop logistics conceptions, I have listed in an appendix the assignments I held in my last twenty years of active service, with indications of some of the assigned or assumed logistic missions of the assignments.

The conception for this paper came from Dr. Hugh M. Cole, Vice President of the Research Analysis Corporation, who had the gracious thought that just as we usually prepare a statement of problems encountered and lessons learned as we close a theater of operations, so I might prepare such a statement as I close my career in logistics. I am indebted to General T. T. Handy, Mr. Conway J. Christianson, and Dr. Charles A. H. Thomson for their review of this paper and their many helpful suggestions.

1970   CARTER B. MAGRUDER

Contents

Recurring Logistic Problems

As I Have Observed Them

CHAPTER 1

Supply Requirements for an Overseas Theater of Operations

The stimuli that activate the logistic supply system in war are the requirements of the theaters of operations. It follows that the first logistic problem area dealt with should be the determination of those supply requirements. The most critical recurring problems that I have observed in that area involve:

1. The determination of resupply requirements for major items of equipment.

2. The determination of ammunition requirements.

3. The use of local resources in a theater of operations to reduce requirements on the United States.

Each of these problems is dealt with separately below.

Determination of Resupply Requirements for Major Items of Equipment

In World War I the U.S. Communications Zone included three echelons of depots: advance, intermediate, and base. Stockage at each was based on French experience. Since the French had three years of wartime experience, these stockages were reasonably adequate. The supply system operated on the assumption that no shortage existed below a depot level unless unfilled requisitions had caused "due-outs" to be established at that depot. Thus a depot requisitioned on the next higher depot for any shortages below its authorized stockage level plus any "due-out" minus any "due-in." The beauty of the system was its simplicity. It worked well in a stabilized theater where the desirable depot stockages had been established by ample experience.

At the time of its entry into World War II, the U.S. Army had no current supply consumption experience. We soon encountered difficulty in computing in the United States the requirements for the resupply of the North African Theater of Operations. The theater had equal difficulty in forecasting its own requirements when requisitioning as far in advance as necessary. In addition, we all had difficulty in determining the desirable level of the theater reserve which should provide protection against errors in forecasting requirements. These difficulties resulted from a lack of understanding of what constituted good replacement factors as well as a lack of data from which to derive them.

In the first major U.S. offensive operation in World War II, American troops landed in North Africa. Brig. Gen. Thomas B. Larkin, who was responsible for the logistic support of the task force, which was made up from our troops in Great Britain and was to land at Oran, came back to Washington to make logistic arrangements. He asked that the War Department provide automatic resupply until the theater was ready to requisition. I was the action officer in charge of arranging the logistic support of the theater. In my previous year on the G–4 War Department General Staff and in the headquarters of the Army Service Forces, I had never even heard of automatic resupply for anything except food, clothing, and POL (petroleum, oil, lubricants). Nevertheless, since it was obviously unreasonable to expect theater service chiefs to prepare requisitions in the confusion of a landing operation, I supported his proposal. I passed the requirement for automatic resupply to the Chiefs of Technical Services, who had knowledge in this field from having had to estimate consumption in preparing their budgets. General Larkin undertook that the theater would requisition for additional quantities of items for which the resupply rates proved too low so that they could be raised and to report items being received in excess of requirements so that resupply rates could be cut. This system worked, although not as well as we had hoped. Under the stress of combat, tactical movement, and a long line of communications, the theater reported shortages but not excesses. In addition, neither of us had recognized the importance of having the theater report such critical logistic actions as the reequipping of the British contingent with American signal equipment and the formation of provisional units equipped from theater reserve stocks. Such withdrawals from depot stocks were reflected only by requisitions on the United States and therefore appeared as normal consumption to the technical services.

After several months, the reported supply consumption was so erratic that a mission was sent to the North African Theater of Operations to determine the causes and to establish replacement factors. Also, the Chiefs of Technical Services in Washington established closer liaison with their service chiefs in the theater. Replacement factors gradually improved but never reached any high degree of accuracy. There were too many complications. The theater troop basis was constantly changing as monthly convoys brought in new increments of troops. Materiel varied in model and age. Troops arrived in the theater with shortages in their initial equipment, both intentional and unintentional. For example, because of the shortage of cargo space, we intentionally shipped the first contingents to North Africa with only half their truck transport. Unintentional shortages occurred when equipment ordered for units failed to be delivered to them before embarkation or was lost en route. Many excesses, however, were never drawn down. This occurred because busy and poorly trained technical service personnel prepared many of their requisitions based entirely on replacement factors without considering current stockage and probable future operations, and partially because there was little disadvantage attached to having too much supply, whereas campaigns could be lost from having too little.

In World War II, "order and shipping time" for most theaters was about four months. This meant that the time between the preparation of a theater monthly requisition and the delivery of the requisitioned items was four months, provided of course that the items were already in stock in the United States. The four-month time lag was equally applicable in automatic resupply. The time lag meant that the theater, or the technical services in the case of automatic supply, also had to estimate the consumption of each item for the next four months and requisition accordingly. These estimates required the use of replacement factors.

Early in World War II the general understanding was that a replacement factor for an item was the percentage of the total quantity of that item in the hands of troops in a theater that should be replaced every month, based on average consumption, and that one replacement factor for an item in one theater would do for all theaters. During World War II we recognized that consumption of items varied with the operations that were being conducted. For example, larger amounts of many items were requisitioned when a major offensive was to take place. The Overseas Supply Divisions at the ports of embarkation, which were charged with editing requisitions from the theaters, were also charged with close liaison with the theaters on planned operations and with making judgments as to appropriate allowances above normal consumption. However, recognition was not given at that time to the desirability of establishing different replacement factors for different types of operations.

Consumption of major items of equipment varies tremendously. As an example, I had occasion during a study at the Research Analysis Corporation (RAC) to examine the consumption of several major items of equipment by the First, Fifth, and Seventh U.S. Armies in Europe in World War II. All three armies were in the same combat posture (attacking); all were fighting the same enemy; and all had a similar balance among combat troops, combat support troops, and service troops. It could therefore be expected that their rates of consumption would be quite uniform. But in looking at the consumption of several selected items by the First Army as a base, the consumption by the other armies of the same items averaged 69 percent for items consumed at rates below First Army rates, and averaged 171 percent for items consumed at rates above First Army rates. These figures show normal variations and indicate that under the most favorable circumstances forecasts of consumption should not be expected to be more accurate than between approximately 69 percent and 171 percent of the actual consumption.

The difficulty of accurate forecasting is well illustrated by the "Report of the General Board, U.S. Forces, European Theater of Operations," prepared immediately after the close of hostilities. It shows that the First Army, even with the advantage of known consumption rates experienced in North Africa, Sicily, and Italy, overestimated its monthly replacement factors very heavily, and the War Department, although much more conservative in its estimates than the First Army, still overestimated quite heavily on twenty-two items, while underestimating on only four.

It is my opinion that if the experience data available are fairly recent and from a similar war, a good replacement factor may forecast consumption with an accuracy between 50 and 200 percent when applied to average consumption over a period of six months or more. Deviations for short periods will be much greater. The degree of accuracy attainable in estimating replacement factors for a war forecasted against an enemy who may have many surprises in store will probably be lower than the 50–200 percent range, with a corresponding degree of danger of critical supply shortages. This danger must be considered in determining the size of the reserves to be established in any theater of operations and in the United States.

Even with good replacement factors based on valid experience, a period of unexpectedly intense operations could cause critical shortages in less than the four months usually required for "order and shipping time." To provide against unexpected requirements, expenditures, or losses is one of the functions of the theater reserve.

The theater reserve is normally expressed in "days of supply." A day of supply is generally understood by the military as a quantitative term indicating a general average of daily consumption. As an example, if an automobile is run on the average 12,000 miles a year, then a 60-day supply of gasoline would be the gasoline to drive the car 2,000 miles, although, with fairly hard driving, 60 days of supply could be exhausted in 4 days. For major items of materiel a day of supply is a function of a replacement factor. Thirty days of supply for any specified item for any theater of operations is computed by multiplying the replacement factor for the specified item by the quantity of the item in the hands of troops in the theater.

The theater reserve in war has three basic purposes. First, it must provide immediate replacement of materiel consumed

above that forecast based on replacement factors. Thus it
offers protection against errors in replacement factors and
against unexpected surges in consumption. Second, it must
tide the theater over initially until the line of communications
with the United States is in full operation and thereafter whenever the line of communications is interrupted. In World War
II, interruptions were caused by submarine attack on convoys.
In World War I, the equivalent of an interruption was caused
by uncontrolled shipments of cargo jamming the railway yards
of New York so that they had to be emptied by shipment
overseas whether the supplies were required by the theater or
not. In World War II and Vietnam, interruptions were caused
by the jamming of overseas ports with cargo above the capacity
of the ports to unload. Third, must provide equipment for
urgently needed provisional units, formed from personnel replacements. In World War II it was extensively so used in
North Africa; in Korea it was used on occasion to provide
equipment for newly formed Korean units.

Toward the end of World War II—with replacement factors
firmly established by recent experience, with theater reserves
generally built up above authorized levels, with convoys arriving regularly without loss of ships, and with theaters so well
established that provisional units were a rarity—a theater reserve of 60 days of supply was reasonably adequate every-
where. Such a reserve level would probably be inadequate for
any active theater at the beginning of a war. Certainly in any
future conventional war in Europe, a 60-day reserve will be
completely inadequate until the nuclear submarine threat has
been overcome and replacement factors have been established
by experience to reflect changes in enemy materiel and tactics.

In preparation for any new war, the best possible forecast
of replacement requirements for materiel for each overseas
theater must be made to guide the laying up of reserves in the
theaters and in the United States and to shape the establishment of industrial production capacities. The initial flow of
supplies to any new theater will inevitably contain excesses and
shortages even if a reasonably accurate operations plan has
been made and if reasonably good replacement factors are
available. The theater reserve should be large enough to take

care of most shortages. The important requirement is to minimize the number of shortages that cannot be filled by the theater reserves. Effective action can readily be taken to expedite shipments from reserves in the United States and, if necessary, to increase production if only a few items are involved. When many shortages exist, the expediting effort is spread thinner and is less effective.

As long as they remained in existence, the Chiefs of Technical Services were charged with maintaining current replacement factors for items for which their services held responsibility. No standard procedure was prescribed for these determinations nor was any documentation required. For several years after World War II, replacement factors from World War II experience were considered adequate. The Army's experience in Korea caused relatively minor changes. There were, however, new items coming into the supply system, such as nuclear weapons and missiles and their launching and control equipment, for which there was no experience. A Materiel Requirement Review Panel was established in 1951 in the Department of the Army under the chairmanship of G-3. The panel had general officer representation from G-4, G-2, and the Comptroller. I was the G-4 representative. The formation of this panel gave recognition to the breadth and the level of interest that ought to be given to materiel requirements. At one time we reviewed the replacement factor for the medium tank and, as I remember it, raised it from 12 to 14. This change was based solely on the improvement that had been made in antitank guns and the start of development of antitank missiles. I realized at the time that many other influences such as tactics and wear-out ought to be considered and that we should be able to attack the problem in small segments, but I did not at the time conceive a solution.

Responsibility for the computation of replacement factors now rests with the Army Materiel commander. In 1964 the Army Materiel Command, having experienced difficulty defending its requested appropriation for procurement of equipment and missiles because of the Department of Defense's contention that replacement factors did not reflect changes in materiel, tactics, and environment, sought the assistance of the

Research Analysis Corporation to develop a sound methodology for computing replacement factors. Such a methodology was developed under a project entitled System for Estimating Materiel Wartime Attrition and Replacement Requirements (SYMWAR). I took part in this project. The first effort of the SYMWAR project was directed toward developing a methodology for estimating replacement factors for a future conventional war, although this focus was justifiably subject to some criticism. The methodology had nothing usable for nuclear war, although war games indicate that NATO could not hold against a full-scale attack by the Warsaw powers without at least battlefield nuclear weapons.

SYMWAR's conventional war methodology provided for variations in the replacement requirements for items of materiel based on changes in environment (as between North Africa and France), types of operations to be conducted, and changes in both U.S. and enemy materiel and tactics. The heart of the methodology was the loss rate table. In most cases, three such tables were required for each item: one derived from the best historical data available, a second derived by updating the first to reflect changes known to have taken place, and a third derived by modifying the second to forecast for a future war by reflecting changes expected to take place between the present and the date of that possible war.

The best historical data for any item then in use usually came from World War II. World War II information was admittedly inaccurate and incomplete, so its use introduced inaccuracies into our computations. The methodology, however, was developed for both current and future use, and the reporting systems now in use in Vietnam and expected to be in use in future wars should provide historical data of greater accuracy.

The loss rates derived from historical data for each specific item were tabulated in a two-dimensional historical loss rate table. This table had a column for each "posture" (attack, defense, withdrawal) with data on such factors as area fire, rocket launcher fire, mines, air attack, damaged and abandoned, wear-out, etc.

We found we needed to give special consideration to the postures mentioned above because materiel consumption appeared to vary widely depending upon which posture a force was in. Experience in World War II with respect to personnel had indicated such wide variation in casualty rates for different postures that personnel casualty rates were included in FM 101–10 for some 14 postures, ranging from Covering and Security Force Action, through 5 variations of the attack posture, through 6 variations of the defense posture, to Pursuit and Retirement and Delaying Action. However, since our materiel replacement factors were to be used in conjunction with a forecast of tactical operations (to provide a basis for requisitioning several months before delivery and to indicate desirable changes in materiel production rates up to eighteen months in advance), we felt that there would be no purpose in having postures broken down to a degree of detail beyond our capacity to forecast tactical operations. Moreover, the materiel interest is in losses over periods of months, not days, as in the personnel casualty interest, and the time that a field army has remained in 1 of the 14 postures related to personnel casualties has rarely been as long as a month.

We did find sufficient justification for using the 3 postures for estimating materiel consumption. A month in the attack posture results in above-average losses of materiel, such as armor, in the forefront of the battle exposed to the direct fire of the enemy. A month in the defense posture results in below-average losses of materiel because materiel is concealed from direct observation and enemy artillery has to resort to area fire, which is not primarily effective against materiel. A month in withdrawal results in very high losses of mechanical equipment because breakdown results in abandonment. Withdrawal also results in almost complete loss of forward stockages of supplies because transportation is seldom available to move more than a few days of supply as fast as the troops move.

We found we needed to give special consideration to cause of loss. Losses of any specific item of materiel are caused principally by the weapons the enemy uses against it and by abandonment, accident, or wear-out. Subdivision of losses in any posture by cause of loss permits the effects of new weapons to be contrasted with the effects of the old. Thus an improved enemy antitank gun will increase tank losses to direct fire, and the degree of its increased effectiveness over its World War II counterpart can be estimated by comparing the loss rate for tanks from direct fire in World War II. On the other hand, the improved antitank gun will not affect wear-out. Such effects as those resulting from improved design must be estimated by referring to the loss rate for tanks from wear-out in World War II.

Loss rate tables can be developed without historical data by analyzing other types of data such as that on exposure to enemy attack, obtained from maneuvers; data on vulnerability to various types of enemy fire, obtained from proving ground tests; and data on enemy and friendly tactics and estimated personnel casualties, obtained from war games. This approach can be used to supplement the historical approach. It was the only method the SYMWAR project developed for new items.

It is my opinion that both approaches should be used whenever data is available. The results should be compared and adjusted by judgment. The historical approach should be worked through whenever historical data is available for an item even remotely similar to the item under consideration because the results in actual combat reflect the effects of so many factors that are indeterminate in a theoretical study.

Finally, a procedure as complex and as heavily dependent on judgment as the one developed in SYMWAR will give good results only if it is kept in continuous use by the AMC staff and the commodity commands, so that replacement factors reflect the latest information and so that the procedure can be improved by those who use it. I also believe that enough of the conception should be taught in the Army's schools so that those who must report the needed data accurately during combat will understand the reason why they are asked to bear this burden and the value of the end result.

As already indicated, the methodology relied heavily on experience data from previous wars, and such data had not been adequately reported and assembled in either World War II or Korea. Then the Research Analysis Corporation was asked to help design and operate a materiel loss reporting system for Vietnam, one could be made to meet the data requirements of the SYMWAR methodology. RAC eagerly accepted, and a reporting system designed by the Army and RAC was installed. Called the Combat Loss and Expenditure Data–Vietnam (COLED–V), it met all the SYMWAR data requirements except for "time in posture," which could be estimated. Thereafter, the SYMWAR project deviated from the original plan and gave primary attention to forecasting requirements for counterinsurgency, even though supplies never have been and probably never will be procured in advance for counterinsurgencies. Because stockages for conventional war are similar to and so much larger than those for counterinsurgency, the assumption has always been made that if the United States laid up reserves for a conventional war in Europe, there would be plenty for any counterinsurgency. This diversion of effort delayed the project, so that now the nuclear war aspect has been dropped.

Reasons why the problem of the determination of resupply requirements for major items of equipment recurs.

The problem of deciding the desirable size of the theater reserves, the CONUS materiel reserve, and the industrial production capacity that should be accepted for budgetary action occurs and recurs because of two conflicting points of view. The military have some recognition of the variations in materiel consumption in the past and of the difficulties of forecasting the future; they intentionally subordinate money to lives and cost to effectiveness; and they always keep in mind General Omar N. Bradley's observation that "in war there is no second prize for the runner-up." Consequently they seek protection in high levels of materiel reserves and production capacity. On the other hand, the civilians in the Department of Defense and the Bureau of the Budget have less conception of the variations that must be expected in consumption of materiel or of the effect of shortages on military operations. They are also closer to popular pressures which make preparations that might help win an ill-defined war in the indeterminate future seem far less important than holding military requirements down in order to leave funds for other government purposes without straining the national economy. They thus direct their efforts toward eliminating the military requirements that they do not consider adequately substantiated. There is so much judgment involved that neither side can prove the other wrong, and the argument is reopened with each new budget request.

Forms in which the problem of the determination of resupply requirements for major items of equipment will probably recur.

When judgment is so large a factor, the military will never be able to convincingly justify any level of supply requirements except by developing replacement factors of increased validity that they can defend with greater assurance and perhaps with some greater degree of success. The forms in which this problem will probably recur will be two: first, the need to determine replacement factors for major items of materiel in conventional war in an accurate and defensible manner and, second, the need to determine replacement factors for major items of materiel in nuclear war.

Determination of Ammunition Requirements

The amount of ammunition required for any operation is not subject to very accurate determination. Generally speaking, the more ammunition that is fired, the easier it is for us to advance and the lower are our casualties. In World War I, after the failure of the Nivelle offensive and its very heavy casualties brought on mutiny in the French Army, efforts were made to restore morale by blasting enemy positions with so much artillery fire that most of the defenders were put out of action by death, wounds, or shock. This permitted enemy positions of limited size to be taken with smaller losses than had been previously incurred in such operations. Messins Ridge, Malmaison, and Verdun (1917) were examples of this type of operation. But neither the availability of artillery units nor the production of ammunition was ever so great in World War I as to permit this type of attack to be used on a broad scale.

In World War II, ammunition was available in relatively less quantity than in World War I. The heaviest bombardment by air and artillery combined was probably that on Cassino. It was nonetheless insufficient to kill, wound, or demoralize most of the defenders. Ammunition in such quantities was not often available. In 1944 ammunition procurement, which was being carried out at rates based on North African and early Italian experience, fell considerably short of demand in Europe, and it sometimes became necessary to ration ammunition severely. For a time during the invasion of Southern France, for example, although we in Italy who were supporting the invasion had allotted the Sixth Army Group its proportionate share of our ammunition supply, the U.S. 3d Division of that force was rationed to ten rounds per gun per day, an amount that could easily be fired in five minutes. Commanders, once committed, seemed to push their troops to about that point where they would accept casualties without undue loss of offensive spirit. Commanders were reluctant, however, to initiate any operation without assurance that ammunition would be available in the quantity the commander felt was necessary. How commanders reached such determinations was an individual process.

In Korea, although our artillery had been firing about twelve rounds to the enemy's one, General Van Fleet complained that there was a shortage of artillery ammunition. A congressional committee held an investigation. The committee established that Congress had appropriated all the funds that the Army had asked for ammunition. The fundamental question, however, was not settled or even clearly defined. It was, What is the proper basis for ammunition requirements? As an artilleryman who had spent four years as an instructor in the Gunnery Department of the Field Artillery School and as a logistician who had been particularly attentive to ammunition requirements in an active theater in World War II, I thought it was the ability to deliver effective fire on all worthwhile targets. These included targets identified as important areas of enemy occupation or activity by such means as visual observation or aerial photographs. They also included, for example, terrain features not known to be occupied but which, if occupied, would hamper an enemy attack. In a defense they also included possible enemy assembly areas. In other words, to me and to most artillerymen, a worthwhile target was one whose destruction or neutralization would make a direct contribution to the success of the mission upon which the command was engaged.

As a Deputy G-4 of the Department of the Army during the congressional investigation, I sent questionnaires to a great number of artillery battalion commanders who had served in Korea, asking each if he had ever been unable, because of a lack of ammunition, to fire on a target that he himself had considered worthwhile. The answers were unanimous in the negative. Opposed to this concept of worthwhile targets was the practice that had been established in Korea of firing the artillery even when no worthwhile targets, under my definition, were available—a practice of firing on areas that had no direct connection with the mission of the engagement or even of firing large amounts of ammunition when our troops were not engaged at all. Artillery had fired harassing and interdiction fires in World Wars I and II, but expenditures for this purpose were strictly limited. The difference in Korea was one of degree: heavy ammunition expenditures in harassing and interdiction fires were not only permitted but insisted upon. As a result of this practice, artillery ammunition expenditure rates being experienced in Vietnam when the supported troops are inactive are about equal to those experienced when the supported troops are in combat.

The principal difference between World War II on the one hand and Korea and Vietnam on the other was that in World War II operations ammunition production was inadequate, and even in late 1944 severe rationing was necessary. In Korea ammunition production was relatively greater, so rationing was much less severe. In Vietnam ammunition production has been ample, so rationing has been unnecessary. In both Korea and Vietnam heavy expenditures of ammunition have been used in the hope of keeping personnel losses down. Ammunition consumption figures from World War II were not applicable in Korea and Vietnam; in the latter cases, the limitation on consumption apparently became more a function of the capability of service units to handle local distribution, of the physical endurance of artillery personnel in serving the guns, and of the ability of the guns to stand such rapid and prolonged firing.

Thus it appears that there are two possible expenditure factors that might be used for ammunition. One would be derived from World War II experience in periods when rationing was not severe. Ammunition requirements would be based on the desirability of accomplishing missions. The World War II rates would of course have to be adjusted for changes in artillery techniques, tactics, and materiel that have taken place since World War II. The second type of expenditure factor would be derived from the Korean and Vietnamese experience, when ammunition supply was virtually unlimited. Ammunition requirements would be based on the desirability of reducing personnel casualties and would be limited primarily by the capability of the artillery to fire.

Reasons why the problem of ammunition requirements recurs.

It has never been politically desirable to determine the value of a casualty. As a result budgetary personnel in the Department of Defense, faced with the desirability of holding down fund expenditures, urge the military to reduce "waste" of ammunition. On the other hand, military personnel, faced with the desirability not only of securing added insurance that missions will be accomplished but also of holding down casualties, urge the provision of all the ammunition engaged forces can fire.

Form in which the problem of ammunition requirements will probably recur.

The form in which the problem recurs is a requirement to establish ammunition expenditure rates, the desirable size of the ammunition reserve, and the ammunition production base. It is probable that no mention ever will be made in any task order of any desirable balance between ammunition expended and casualties suffered. There is now evidence, in the congressional action on the 1969 budget, of an unwillingness to appropriate for Vietnam all the funds requested. This reluctance is normal in any postwar period or period of peace. It will force the Army to give more consideration to priorities within its own requirements as it seeks to secure the greatest possible military strength for the dollars available both for the current war and for any future war.

Regardless of the funds available, in any major effort the military services cannot have everything they want. The availability of industrial manpower, of production facilities, and of natural resources is limited. In World War II the sum of the requirements of the military services far exceeded the productive capacity of the United States, and so the requirements had to be reduced. Even then, as I remember one writer putting it, we consumed the Mesabi Range on the battlefields of Europe. Accordingly, the Army and the Department of Defense must recognize that more conventional ammunition means fewer tanks, fewer bombs, and fewer ships or less of some other related critical requirement. Similarly, more battlefield nuclear weapons mean fewer intercontinental ballistic missiles, fewer Polaris or Poseidon missiles, and fewer aircraft or less of some other critical related requirement. These considerations will probably cause the problem to recur in the form of a requirement initiated by the Army itself to establish to what degree the heavy ammunition expenditures in Vietnam are justified.

Use of Local Resources in a Theater of Operations To Reduce Requirements on the United States

The United States has a great wealth of resources and a history of being the "Arsenal of Democracy." In addition, Americans in general have a feeling that everything good is produced in the United States and that the products of other countries are strictly second-rate. As a result, American overseas theaters in war are inclined toward requisitioning on the United States to the maximum extent. The British, on the other hand, lacking resources at home and used to seeking them in colonies, were much better prepared in World War II to live off the country to whatever extent was practicable.

Local procurement overseas has many advantages. It saves our resources. It saves time and shipping space. The use of local labor reduces the requirement for U.S. service support troops. If the local construction industry is used, local procurement provides equipment and materials with which local workers are familiar. Local procurement overseas makes it possible for a supported nation to make a greater contribution to the common cause or a conquered territory to be forced to pay a part of the cost of the war.

I have encountered many illustrations of the advantages to the United States of local procurement overseas. Most of these illustrations were in World War II, because local resources in many of our overseas theaters in that war were abundant, but there were some in the wars in Korea and Vietnam. In all three wars the use of local labor was of major importance in reducing requirements for American logistic troops. Also, since local labor usually derives most of its own logistic support from the local economy, there has been a considerable saving in support tonnage required from the United States.

I can illustrate most of the advantages of offshore procurement by an example of railway operations in Italy in World War II. Generally speaking, all the Italian railways were under the control of our Military Railway Service, an integrated British-American organization. The British and American railway troops, however, only had to handle rehabilitation and operation in the forward areas because they could turn over the operation of most of the lines in the rear to the Italians. In withdrawing in 1944 to a position covering Bologna, the Germans had been thorough in destroying the two railway lines, one on each side of the Arno River, that connect Leghorn with Florence. Every bridge was demolished, every tunnel blown in several places, every culvert blown, even every single rail cut. In preparing to support the American Fifth Army in the Po Valley campaign in the spring of 1945, it was necessary to put one of these railway lines back into operation.

I was G–4 of the theater. Our original plan was to restore the rail line north of the Arno where reopening the Serravale Tunnel appeared to be less of a problem than replacing the ninety-six bridges and culverts that would be required on the rail line south of the Arno, especially since the South African Mines Brigade was made available for tunnel repair. Three months before the spring offensive was due to commence, it became apparent that a stratum of mud, in one of the places in the middle of the mountain where the tunnel had been blown, was causing so much difficulty that the tunnel could not be restored in time. The restoration of the railway line south of the Arno then became a very high-priority project. Brick and mortar were available locally as were Italians skilled in masonry, so contracts with Italian industry could be used to repair culverts or to replace temporary bridging with permanent masonry. The rebuilding of bridges, of which six were between 80 and 100 feet long and one was 500 feet long, however, was too urgent and difficult to attempt by contract. Two battalions of American engineers and the construction company of a railway operating battalion were charged with the bridge building. The project would have been impossible to execute if we had been forced to requisition on the United States for railway bridging. Fortunately, the British had put an Italian firm (my memory is that it was the Tierne Steel Works) into operation, producing prefabricated steel railway bridge piers, bridge spans, and girders. These were made available as soon as we could use them, and the rail line was reopened before the Po Valley campaign commenced.

I assume that Italy ultimately paid for most of this Italian assistance since under the armistice Italy undertook to make available such facilities, resources, and services as our forces required. The Italian government was not at the time, however, able to reimburse contractors and vendors furnishing facilities, resources, and services to the Allied Forces on requisition as was done in France. When we needed Italian currency in order to carry out local procurement, our finance officers drew lira from the Allied Financial Agency. These lira were accounted for just like dollars, with final settlement of accounts to be made at the peace table.

Payment for the local procurement of facilities, resources, and services was a complicated and difficult problem that seemed to be handled differently in every country in which we operated and sometimes by several methods in one country, all because of differing international agreements. We started in North Africa with dollars on which the U.S. Treasury seal was printed in yellow (and usable in the United States as well as in North Africa) and ended up in Germany with marks, provided from the Termination of the War (TOW) budget and were handled just like our appropriated funds in the United States. Although all methods provided the essential ability to accomplish local procurement, I preferred the TOW budget—even though I found it just as hard to defend the occupation forces' TOW budget requirements to General Clay as I ever did later to defend the Army's procurement budget to the Appropriations Committee of the Congress. The use of the regular appropriated fund procedures, which were well understood by the American military, protected the German economy from unreasonable demands and provided better supervision and control over American personnel handling TOW appropriated funds than we had been able to exercise under the systems previously used.

Although the use of occupation marks was an excellent way to place some of the cost of the occupation on the Germans, we had one major problem with our marks. In addition to using marks issued by the occupation authorities for local procurement, we also used them to pay our American military personnel. However, we allowed American military personnel to buy, for these occupation marks, postal money orders in dollars for transmittal to the United States. The Soviets also paid their personnel in the same marks and in large amounts. With the shortage of goods and the excessive availability of marks, the mark depreciated. Some Americans traded goods, either procured from the post exchanges or ordered from the United States, for depreciated marks and then converted large quantities of these marks into dollars through postal money orders. We stopped this practice by a restriction which in essence terminated any conversion of marks into dollars. Military personnel wishing to send dollars to the United States could increase their allotments, have pay checks deposited in American banks, or buy money orders at the pay table.

Local procurement is advantageous not only to the military forces, but also to the local economy. Procurement in a theater of operations helps to keep the local economy operating. I can illustrate this with the very smallest of examples. When I had the IX Corps in Korea in 1954, I noticed our soldiers fishing in an irrigation reservoir but never catching anything. I dispatched a young captain to find a fish hatchery. He returned with the information that there appeared to be only two still active in Korea. They produced only one type of fish called the red horse, a species of carp. The Koreans used the red horse as a food fish. They released them as fingerlings in the rice paddies when they were flooded. The fish grew to eating size by the time the rice paddies were ready to be drained in the fall. In the disruption caused by the war, the demand for red horse had disappeared. As the only customer in sight, we contracted for the whole output of one of the hatcheries. I was ordered away before the fishing season, but I hope our purchase helped to keep the hatchery going until the commercial demand reappeared.

In my opinion, U.S. military procurement, with its requirement that specifications be met, gave great impetus to quality control in Japan during the Korean War and helped change Japan from a nation that was noted for making cheap but poor copies of the products of other nations to a nation whose products were competitive in quality. Similarly, I believe that some of the current improvement of the Korean economy can be traced to local U.S. military procurement since the Korean War. During my period of command, I made every effort to encourage the Korean economy to produce for both the military and personal requirements of the American forces. One of our more successful efforts was what Maj. Gen. James A. Richardson called his "County Fair." This fair consisted of two buildings whose walls were covered with items, mostly hardware, that we thought the Koreans could produce and that we wanted to buy. By each item was a card showing how many we had bought in previous years, how much we had paid, and how many we expected to buy in the current year. With the exhibits were American experts who could discuss manufacturing processes with any prospective supplier. In our procurement we required the Korean products to meet American standards. Our exchange service made special efforts to procure Korean products for resale. Our quartermaster sought to buy Korean food products that met our standards of quality and sanitation. At first only the produce from the Seventh Day Adventist mission met these standards, but gradually more and more Korean food products did. One of our particularly successful steps forward was with the Korean fishing industry, to which we lent some of our veterinary personnel to teach sanitation and from which afterward we were able to buy seafood in considerable quantity.

The disadvantages of local procurement are few. On occasion, it may strain the local economy. (Thus in 1946, with food critically short in Germany, we stopped local procurement there of food for the American occupation forces.) More recently, offshore procurement has been restricted severely to cut down the gold flow in spite of the increased cost of importation from the United States.

Our local procurement efforts in the past have been primarily the result of the initiative of the military personnel of the theaters of operations rather than the result of any guidance from the United States, primarily because those in the field could best see the possibilities at hand. Procurement efforts have been directed primarily toward big-tonnage items because the military authorities overseas were thoroughly alert to the criticality of transportation. The other incentives for local procurement have been time and convenience. No accurate information has been furnished overseas theaters on the needs or shortages of the United States economy. Some scrap iron, iron ore, and phosphates, for example, were shipped home from North Africa in World War II primarily as ballast. Much more could have been usefully shipped, although at some cost in loading time. On the other hand, a large quantity of oranges shipped to Britain from North Africa at the initiative of the theater stirred up a protest from importers that was so violent as to completely overshadow any expression of appreciation from the recipients who rarely enjoyed fresh fruit during the war.

In the Korean War, some facilities, some construction materials, and a great amount of local labor were obtained locally. The mass of offshore procurement, however, was accomplished in Japan. The most spectacular example of local procurement was the tremendous operation established in Japan to rebuild American World War II equipment gathered from the whole Pacific area. One of the most important aspects in this operation was the capability of the Japanese to produce needed repair parts that were out of production in the United States. Japanese industry had so much experience copying western products that "reverse engineering" of a part for which no specifications were available seemed to give little trouble. This rebuilding capability in Japan was used after the Korean War not only to support U.S. and Korean forces in Korea but also to process military aid equipment for other Far Eastern nations. It has been used extensively in support of the war in Vietnam.

Reasons why the problem of the use of local resources recurs.

As long as the gold flow from the United States needs to be reduced, the reduction of local procurement overseas will continue to be studied. On the other hand, once the gold flow is corrected and a new war threatens, recognition will probably be given to the fact that U.S. resources are not inexhaustible. Impetus will then be given to increasing procurement overseas. Also, there may well be a long period during which the United States must retain troops in Vietnam and bolster the Vietnamese economy. During such a period, American military personnel would provide the equivalent of both an export market and a tourist trade. Local military procurement, both for the U.S. forces still in Vietnam and for military aid being provided by the United States to the Vietnamese forces, could be of great value in developing and supporting the local economy.

Form in which the problem of the use of local resources will probably recur.

If war ever threatens in Europe, maximizing local procurement overseas will become critical to reduce shipping requirements, given the threat of the nuclear submarine. Also for such a war and for any other large war that may threaten, the necessity of conserving U.S. resources should lead to studies to determine, in advance, what items that are short in the United States can be procured locally and what monetary or requisitioning process should be used to procure them. The problem will also likely recur in the form of studies of how to foster the Vietnamese economy during the recovery period.

CHAPTER 2

Logistic Troop Requirements

Logistic troop requirements usually are, and I believe always should be, under attack. This is for the simple reason that more personnel in logistic units means less personnel in combat units. However, after a critical review has been made and logistic troop requirements for an operation adequately justified, then those logistic troops should be considered just as essential to the success of the operation as the combat troops. Although most commanders would probably agree to such a statement, it just never seems to be handled that way. Rather, it is handled as a logistician in the North African Theater of Operations described a troop basis meeting in early 1943. He said new troop units were asked for on the basis of parity: one tactical unit and then, in turn, one logistic unit—except that the approved list came out in the ratio of one infantry division to one heavy maintenance company.

The most critical recurring problems that I have observed in the area of logistic troop requirements are:

1. The shortage of logistics troops at the beginning of a war.

2. The need to determine measures, and their price, that can reduce the requirements for American logistic troops in a theater of operations.

Shortage of Logistic Troops at the Start of a War

Military forces require much the same type of services as the civilian community. As a result, most of the special skills needed already usually exist in any large group of mature personnel brought in by the draft. Such skilled personnel need only to be taught the military aspects of their skills and to be

trained enough to acquire a minimum amount of discipline in
order to form satisfactory logistic troops. Combat troops, on
the other hand, must be taught skills not developed in the
civilian community. They need to be toughened physically to
endure the much greater rigors to which they will be exposed.
They must acquire a much higher level of discipline to hold
them to their tasks under hardship and danger. Because of
these requirements, it is generally understood that it takes a
year to train a combat unit but only six months to train a
logistic unit.

In time of peace the most basic control on the military is
the limitation on personnel strength. In order to get the maximum military strength with a limited personnel base, the
longer training-time requirement for combat units dictates that
most of the units of the regular establishment must be combat
units. Logistic personnel on active duty in peacetime are generally limited to those required to provide peacetime logistical
support for the active combat forces. Combat forces require far
less logistic support in peace than in war. In peace, shortages
of materiel do not have to be replaced promptly; construction
can be done by contract; there are few sick and no wounded to
care for; and maintenance requirements are light because
equipment is less used, better cared for, and not damaged by
enemy action.

When combat troops move out of their peacetime posts to
go to war, these posts are used to accommodate newly formed
combat units, which require trained logistic support as soon as
they are formed. The departure of the original combat units
does not therefore release the logistic personnel who supported them. In 1939 the 3d Division, then stationed at Fort Lewis,
Washington, left that post to engage in a landing on the
California coast, followed by several months of field maneuvers. This was one of the most important peacetime exercises
the Army has ever held, and it was important that every available man take part. Yet we found it necessary to leave behind
approximately one thousand soldiers from the division to take
care of the post even though it was not occupied by other
combat troops.

Logistic troops in the peacetime troop base are always
vulnerable to elimination. Reduction of logistic troops is called

"cutting out the fat" in press releases. This has been going on in the U.S. European Command in recent years. It took place after World War I to such an extent that at the beginning of World War II the active Army had only 11 percent of the logistic troops necessary to support overseas the available combat units. Lacking sufficient active-duty logistic troops to meet requirements at the beginning of a war, the Army has come to rely on reserve logistic units. A good reserve unit can be called up and put in shape for deployment to an overseas theater in perhaps three months.

Although the Army carries few logistic troops in its active peacetime establishment and has to train reserve logistic units before they are ready to be committed to an active theater, the need for logistic troops precedes the need for combat troops in war. For example, early in World War II we sent combat troops to the British Isles. Supplies were shipped concurrently with the combat troops, but logistic troops to receive, store, and issue these supplies were either unavailable or untrained. It was hoped that, while still training on the job, the logistic troops would gradually identify, inventory, and store the supplies sent. Yet over six months later, when some of these same combat troops were sent from the British Isles to invade North Africa, they could not be equipped and supported from the supplies known to have been shipped to the British Isles. Some of the equipment and all of the supply support had to be shipped from the United States. Thus at a time when equipment and supplies, as well as shipping, were critically short, great quantities of supplies on hand were unavailable for use because trained logistic troops had not been available in numbers to balance the combat troops sent overseas.

A related illustration of the same difficulty in the same war was the accumulation of shipping awaiting unloading in many overseas ports for lack of logistic troops to unload, segregate, and store the cargo. Noumea, New Caledonia, was the most publicized example, with over eighty ships waiting to unload in September 1942. It took four months to bring this backlog of ships down to an acceptable level.

In the Korean War the shipping situation was ameliorated by the availability of the fine Port of Pusan and by the use of

Japan for storage and transshipping by LSTs. The shortage of
logistic troops was a little less critical than in World War II
because the logistic troops who had supported the American
occupation of Japan were available. These logistic troops, however, had to be split between the U.S. troops in Korea and the
U.S. base in Japan. As a result, there were problems in receiving, storing, and distributing supplies in Korea, problems
made doubly difficult because the Koreans had few logistic
troops of their own. At the same time, our first efforts were
mainly directed at developing among the Koreans the capability to organize and train combat units.

It was not until I came to Korea as the United Nations
commander in 1959, six years after the Armistice, that any real
effort was made to establish an all-Korean line of communications to support the Korean Army. This time lag was not
inadvertent. There are many preliminary requirements that
must be met before an effective line of communications can be
established by any foreign country using U.S. materiel. They
include, among other things, a requirement that the foreign
personnel learn English so they can read our manuals or that
our manuals be translated into the foreign language. Neither is
an easy task because of the many technical terms involved.

A still more difficult requirement in many countries is the
inculcation of integrity, which is essential to all military operations but peculiarly to logistics. Integrity emerged as a problem in the Korean Army during my period as United Nations
commander. The Korean Military Academy, patterned on West
Point, had been established during the Korean War. Its graduates were therefore the junior officers of the Army. They had
been taught high standards of integrity, and so, on occasion,
they considered unethical some of the things they were directed to do. I had observed the same problem in much milder
form in our own Army when our post–World War I West Point
classes came in under the integrated World War I officers,
many of whom had lower standards. Some of the latter, for
example, had attended colleges where cheating was normal. I
still remember the protection we juniors felt because we could
call for an inspector general. This assurance that any complaint
would receive a fair and impartial investigation by its very
existence exerted an influence so strong that it rarely needed

to be invoked. Although the Korean Army, had been patterned on the United States Army, it had no inspector general. I urged the establishment of an Inspector General's Department on the Korean Army, and the problem, if not eliminated, was certainly ameliorated. I felt that elimination of the problem could come only when graduates of the Korean Military Academy had reached the top levels of the Army.

One of the principal reasons given for American reluctance to establish a Korean-manned line of communications, pilferage, is worthy of note. Pilferage takes place in every theater of operations. It cannot be stopped under our laws and practices, any more than shoplifting can be stopped, except by an uneconomical amount of effort. What pilferage does take place is usually militarily inconsequential, but publicity makes it a political issue out of all proportion to its true importance. As a result, the military provide excessive protection. For example, I found a company of eighty American soldiers protecting our main Exchange Service warehouse in Korea even as we had to use Koreans to keep our American combat units up to strength. I directed the Exchange Service to contract for Korean guards. This proved satisfactory because when property was lost the contractor, unhampered by American practices, fired guards without having to prove guilt or negligence. The Korean Army proved entirely capable of holding down pilferage to a reasonable level in operating its own line of communications.

Vietnam, in its early phases, showed the effect of a shortage of logistic troops similar to that of World War II. The French had never organized and trained all-Vietnamese units but had furnished officers, noncommissioned officers, and specialists. As a result, when the French withdrew, the burden of furnishing logistic units fell heavily on the United States. Ships awaiting unloading accumulated at Saigon. Great quantities of unidentified supplies accumulated ashore. The situation was not relieved until the United States built and manned a new line of communications, which included a new port and the base depots at Cam Ranh Bay. Many of the unidentified supplies, and some of the accumulated supplies that proved to be excess, were shipped from Vietnam to Okinawa. The identification and inventorying of the remaining supplies are still incomplete (mid-1969). This situation was the result not only of the lack of logistic troops in the regular U.S. troop basis but also of the presidential decision against the calling up of reserve units upon which secondary reliance had been placed.

Reasons why the problem of a shortage of logistic troops at the beginning of a war recurs.

With manpower ceilings being a basic control of the Army, there will always be competition between the requirements for combat troops and logistic troops. Because the combat troops take longer to train and have much more popular appeal, the balance will normally be weighted in favor of the combat troops. Those bearing responsibility for logistic support must continually strive to secure a reasonable balance.

Form in which the problem of a shortage of logistic troops at the beginning of a war will probably recur.

Efforts to support the establishment of a level of logistic troops in balance with combat troops will usually take the form of developing the troop basis for a projected operation in a way that presents the logistic troop requirement in the most convincing manner available. Initially this can probably best be done by confirming or establishing the capacity of each type of logistic unit, determining the tasks to be performed and the time available for each task, and then computing the logistic units required to perform the tasks. This method, however, has been used before without achieving an increased authorization of logistic troops. I believe that in the future the Army will prove the need for logistic troops by using war-gaming to forecast the risks that accompany a shortage of logistic troops. The original concept of the RAC's project on this subject included the establishment through tactical-logistic war games of supply shortages, of their influence on the capability of combat troops, and of the resulting likely effect on the progress of the campaign. This concept has never been carried out, but it will probably be considered again when some responsible logistician, having done everything he can to reduce logistic troop requirements and increase availability, still feels a critical need to support a level of logistic troops higher than that which has been approved.

Measures To Reduce the Requirements for
Theater Logistic Troops

Since efforts to secure what is considered to be an adequate level of logistic troops have usually failed, continuous study of the expedients that have been used or might be used to reduce this need is indicated. Among the possible measures that should be considered are those which reduce the need for logistic troops in general and for American logistic troops in particular. Among these measures are simplification of distribution, improvement in reliability and durability of equipment, reduction in fuel consumption, use of local labor to perform the desired services, reduction of the theater requirement for logistic support by increased use of transportation, modification of maintenance policy to reduce the overseas maintenance load, and organization and training of allied logistic troops to replace American.

Combination of items into packages, so that a relatively small number of packages instead of many individual items can be distributed, has been tried many times with some success in reducing the requirement for logistic personnel. An example of one of the most successful efforts is the K-ration. One package contains one meal for one combat soldier, and individual packages can be packed in quantity in large containers. Thus several cases of K-rations can be supplied a unit instead of the many different items needed if cooks had to prepare foods by combining the necessary ingredients. The disadvantage is that our standard of living in peacetime is so high that merely providing sufficient nutrition is not enough to satisfy our soldiers. Even the greater variety of the C-ration appears unacceptable for periods of long duration. Industry is making efforts along the same line with prepared food mixes. More popular among soldiers, these mixes save some labor and still simplify distribution. This solution is less efficient than combat rations, and in some cases the mixes do not keep as well in storage.

Industry has also created kits that include all the items necessary to make the more common repairs needed by some mechanical items. Thus a repair kit for a specified make and model of carburetor can be purchased. This simplifies distribution but increases supply because all the items in a repair kit are seldom used.

The ultimate military objective is of course a standard container with a standard content that will supply a specified unit for a specified time period. The overseas shipping container developed at the end of World War II was the best size of container for fitting into the holds of ships, into European freight cars, or into military trucks. We worked with standard lists of contents for the occupation forces in Germany but found peacetime requirements too different from wartime requirements to permit useful testing. These containers are still in use and protect their contents against damage in shipment and against pilferage. A disadvantage is the need for a crane in handling them. This disadvantage may be eliminated by the use of the trailers now being carried on flatcars by American railroads and on ships built to carry them. The development of a satisfactory standard content still eludes us. Like the automatic resupply of a theater of operations, almost any standard content will supply many items in greater quantities than required and many, many others in inadequate quantities, if at all. Experience has established that only food is consumed at a uniform rate. Progress is probably possible if the minimum requirements are listed on requisition.

Although improvement in reliability and durability of equipment has always been an objective, we have made only moderate progress. Industry has a limited incentive because so much profit is made in maintenance and repair parts. I have seen only one proposal that I thought might stimulate real action on the part of industry. It was a proposal of Lear Siegler, a manufacturer of aircraft and automatic parts to put in one fixed-price contract, to be awarded competitively, the provision of a specified number of end items and an undertaking to maintain them for their specified life at the expense of the manufacturer. Such a contract would give the manufacturer a strong incentive to build reliability and durability into his product. So far as I know this approach has not yet been tried.

The tremendous fuel requirements of a theater of operations in this motor age invite attention to any possibility of reduction. With the reduction of fuel requirements goes the reduction in the requirements for the construction and operation of pipelines, the erection of tankage, and all the other actions involved in the receipt, storage, testing and distribution of POL.

When I became Deputy Chief of Staff for Logistics, I raised the question of why we still put gasoline engines instead of diesels in our tanks. I wanted the reduced fuel requirement, the greater range, and the reduced fire hazard that the diesel offered. I was told there was a ruling of the Munitions Board based on the belief that the Air Force and Navy needed all the diesel oil that could be secured by the military services. An investigation showed that this was no longer true, so we put a diesel engine in the M60 tank.

The ultimate fuel saving in sight when I was Deputy for Logistics could come from the introduction of nuclear power. Looking forward to the time when our surface-to-air missiles and our antiballistic missiles would provide a considerable measure of protection against intercontinental nuclear attack, I believe that the Army could make an important contribution to combat power in nuclear war if it had armored vehicles that fired nuclear weapons and were nuclear shielded and nuclear propelled. We already had nuclear weapons suitable for mounting in tanks, our tank armor already gave a reasonable degree of protection against the blast and heat of a nuclear burst, though not enough against radiation, and we already had a project for a very large nuclear-powered tractor for a so-called logistic train. I personally did not think much of the logistic train, but I supported the project strongly because I wanted progress toward a small nuclear-powered engine for tanks. The project has since been discontinued, but I look to see small nuclear-powered engines developed when the state of the art is further advanced.

There are many ways of utilizing local labor. If the economy of the country in which operations are to be conducted is operating, it may be possible to contract for services such as the construction of depots, the operation of manufacturing plants or the running of railroads. Varying amounts of assistance and supervision are required. This method was used on occasion in Italy and France in World War II.

If the local economy is in bad shape, a greater degree of assistance is required tending toward more supervision, the provision of some or all equipment and the provision of some or all materials. Furthest down the scale in this area is the Type B unit which is really an American logistic unit with its normal equipment as carried in the Table of Organization and Equipment, but from which all personnel have been removed except those for administration and those possessing certain critical skills. The American personnel withdrawn are replaced with indigenous personnel. Variations of this solution were used in North Africa in World War II and in the Korean War.

Depending on the type of work required, the terms of surrender, the international agreements in force and the political situation, it is sometimes possible to utilize prisoners of war for local labor. Italian prisoners of war who volunteered for the work were used with limited success in North Africa in World War II, as authorized by the Geneva Convention. After Italy capitulated and then declared war on Germany, it was possible to keep the former Italian POWs as Italian military personnel, but it was considered politically desirable to replace many of them with Italian civilians. Also civilians required less support from us than did POWs.

When the German forces in Italy surrendered in May 1945, special terms in the surrender allowed us to use them for a wide variety of military purposes and, as I remember it, regardless of whether or not they volunteered. They were, however, remarkably willing. We used German logistic units in their original organizations under their own officers. They required little instruction, guarding, or supervision. They gave fine service. The Italian public, however, had become antagonistic to the Germans and resented the relative freedom from restraint with which we allowed the German POWs to operate. Although the German POWs were conducting themselves in an exemplary fashion, it became politically desirable to restrict such of our uses of them that puts them in contact with the Italian civil population. In Germany after World War II, we organized Polish and Baltic displaced persons, mostly POWs who had been held by the Germans, into logistic units under their own officers. They too performed fine service. No effort was made to utilize North Korean prisoners of war in the Korean War because great numbers of willing South Koreans were available and needed some means of earning a living.

All the methods of utilizing personnel available locally are dependent on economic and political conditions that are hard to foresee and are usually handicapped by language difficulties. Most required a considerable amount of time before they become effective. Accordingly, primary dependence in the early stages of an operation is best placed in American logistic units with local personnel being utilized as early as practicable.

Where transportation is available, a short evacuation policy reduces the need both of medical units and of hospital construction in a theater of operations. In World War II some of the burden of care for sick and wounded was taken off the theater by the use of hospital ships. In Korea, enough air transportation was available to ship many sick and wounded back to Japan. In Vietnam, air transportation is used still more widely to bring sick and wounded back to Japan, Hawaii, and the United States. A short evacuation policy provides better medical service. It has the disadvantage that many experienced men do not go back to their units and those who do are away longer. This in turn increases the requirement for replacement and lowers the level of experience in units.

Where transportation is available, the requirement for logistic support in a theater of operations also can be reduced by transferring some of the maintenance load to a base outside the theater or to the United States. This was not practicable in World War II because cargo shipping was so critically short. It was extensively used in Korea where ample shipping, particularly LSTs, permitted easy loading for back haul to Japan. It is still more extensively used in Vietnam, with unserviceable items of all types being outshipped to both Japan and the United States for repair. Replacement items have to be shipped in, increasing the transportation workload. The quality of maintenance is probably considerably higher but the repair cycle is much longer requiring a larger supply of end items.

The Air Force has long utilized transportation to reduce maintenance performed overseas. Unserviceable aircraft engines are shipped back to the United States, usually by air

freight, for repair. This has the advantage that better maintenance is accomplished with a lesser stock of tools and parts. It
has the disadvantage of requiring a larger pipe line of engines
and of using transportation that is critically needed for other
purposes in the early stages of an operation. The desirability
of this kind of procedure depends on the weight, size, and
complexity of the item to be repaired. It is usually inappropri-
ate for heavy items like armored vehicles but it is probably
efficient for many light items such as electronic equipment.

Where air transportation is available and a ground line of
communications is difficult to establish, maintain, or protect,
an air line may be established. An air line of communications
may reduce the need of many types of logistic support if the
requirements for the construction, maintenance, and operation
of the necessary air bases are less than the requirements of
that part of the ground line of communications that is replaced. Up to the present, air has been only supplementary to
a ground line of communications, unless the "Dump Run"
across the Himalayas from Chabua in India to Kunming in
China in World War II for the period up until the Burma Road
was reopened can be considered an exception. Recently a
study that estimated the savings of logistic troops that could be
effected if an air line of communications replaced the ground
line was used to help justify the development of the C–5A
aircraft. Such a line of communications has the disadvantage
that a defeat leaves the field army with no logistic base to fall
back upon.

The more end items or parts are classified "throwaway"
instead of "repairable," the less maintenance support is required. In World War II the shortage of logistic units influenced local decisions not to repair many items and parts even
though they were not classified as "throwaway." Maintenance
effort was usually available only to replace major assemblies,
not to repair them. As a result engines, transmissions, rear
axles, and many types of electrical and electronics components,
although not classified as throwaway, were replaced rather
than repaired. In Korea the soldiers often made the decision
for us as to whether or not to repair with respect to end items
normally man-carried. In the rugged Korean terrain any man-

carried load was a burden. Soldiers were unwilling to carry unserviceable items in action and so often threw them away.

For some items, lack of repair parts has at times made the decision for us as to whether or not to repair. Thus in the engineer depot supporting the IX Corps in Korea, I once counted over 120 different sizes, makes, and models of generators parked in outside storage and unrepairable because of lack of parts. They might just as well have been abandoned since the supply of parts for nonstandard equipment is too complicated to be efficiently carried out in war. Increased degrees of standardization reduce requirements for both maintenance and supply units. This subject is covered at length in Chapter 6, "Maintenance of Materiel." But the opposite side of the picture must also be considered. Many actions toward reducing repair increase requirements for supply.

Since we have in the past and expect in the future to operate with allies, and since being a great industrial nation and the "Arsenal of Democracy" we usually supply most of the materiel, we are inclined to assume the very heavy logistic burden of operating a line of communications with our own personnel even though it serves allies as well as Americans. Thus Americans operated the line of communications in the invasion of Southern France in World War II even though it supported a French Army as well as an American Army. Americans operated the line of communications in Korea supporting the Koreans and the troops of many other nations. We do the same in Vietnam. I do not say this is wrong because it is certainly more efficient. I do feel that we should reduce the magnitude of the burden by arranging to start at the earliest practicable date the organization and training of allied logistic troops and turning over to them whatever tasks they can effectively handle. The earliest practicable date could well not await the beginning of a war but be the date on which we start a military aid program.

Reasons why the problem of reducing American logistic troop requirements for a theater of operations recurs.

There is little hope of ever securing in peacetime the activation and training of all the logistic units that will be needed in war. Those responsible for logistic support must therefore take all practicable measures to reduce the requirement for logistic troops and the impact of any expected shortage in such troops in war.

Form in which the problem of reducing American logistic troop requirements for a theater of operations will probably recur.

The problem of how to reduce the requirements for logistic troops will probably recur in the form of studies, research, tests, and plans for direct action oriented toward:

a. Simplification of distribution by broader use of containers with standardized content.

b. Improved reliability and durability of equipment.

c. Use of local labor.

d. Use of transportation to support shorter evacuation policies, to return equipment to the United States for repair, or to reduce ground lines of communications.

f. Reduction of the maintenance load by making more components and parts "throwaway" instead of "repairable."

g. Organization and training of allied logistic troops.

CHAPTER 3

Logistic Support of Contingency Plans

Ever since the armies of Genghis Khan swept across most of Asia and Europe, virtually without any logistic drag on their freedom of maneuver, the progress of civilization has been increasing the quantity and complexity of military materiel and the standard of living of military personnel. As a result, the logistic "tail" of any major force has become so great that the establishment, maintenance, and protection of a line of communications must be assured before any maneuver can be seriously considered. In World War II, when a new theater of operations was being opened, it was a function of Headquarters, Army Service Forces, to recommend, among other things, the logistic units that should operate the line of communications, the level of supply reserves that should be stored in the communications zone, and the provision of any special equipment or supplies not carried in Tables of Organization and Equipment (TO&E). Decisions in these areas determined the size of the logistic tail.

No logistics doctrine handed down from World War I pointed out the importance of special equipment or construction materials. This was because these items had been determined before the United States entered that war. Determining such requirements was therefore a relatively new field for U.S. logisticians when American forces first went into new theaters in World War II. Examples of the more important types of major items of equipment not carried on Tables of Organization and Equipment, and therefore not automatically furnished with troop units, are landing craft and railway rolling stock. Examples of the more important types of construction materials are barbed wire, pierced steel plank, and invasion-type pipe for POL pipe lines.

Troop units carry with them the minimum equipment essential to combat. They do not carry much for logistic purposes. For some of the more important and common logistic equipment, logistic troop units have been formed to provide both the equipment and the operators. Special logistic equipment, needed because of specific conditions in a theater of operations, however, is not necessarily carried in the tables of organization of logistic units. For example, to reduce the requirement for cargo ships in the North African operations in World War II, general-purpose 2½-ton trucks were packed partially disassembled in what was called a "twin unit pack." Assembling these trucks without special equipment was a laborious job. Finally, a troop unit equipped with special equipment and organized and trained under General Motors advisers was sent to Casablanca. But meanwhile, at Oran, improvised equipment had been built from what was available locally, and local laborers had been trained to do the work under U.S. Ordnance supervision. Both assembly plants served the purpose.

It is the usual practice in military planning to prepare operational plans first and then prepare logistical plans to support them. In order to avoid a lengthy "cut-and-try" approach to an acceptable combined operational-logistic plan, we found in World War II planning that an operational concept should be immediately followed by a transportation capability study. Only if the transportation system will support, or can be made to support, the forces necessary to carry out the operations plan is it worthwhile to go farther in the logistic planning.

During the preliminary consideration of proposed contingency plans in the past, I had occasion to demonstrate that certain operations were infeasible because of transportation alone and so was able to avoid all other logistic computations. Examples of such contingency plans included a landing at Genoa in World War II, infeasible because, unless the enemy's transportation system was crippled by a preliminary major air offensive, he could build up forces in the objective area faster than we could; an invasion of Europe through Greece, again in World War II, when the western Mediterranean was still closed, infeasible because the available shipping could not transport and support more than half the necessary force in the objective area; and a plan considered in peacetime, to be implemented in the event of war, for a major force to be landed far inland by air and thereafter supported by air using amphibians (this plan would have given support to the completion of the development and procurement of the MARS heavy amphibious transport plane), infeasible because of the tremendous tonnage that would have had to be moved over a long and vulnerable air line of communications and the resulting size of the required air fleet.

No one who has not carefully studied the subject is likely even to imagine the shrinkage that takes place in a really long line of communications. It was forcefully brought home to me by a chart that General George C. Marshall, the Army's Chief of Staff, called "the intestinal tract." It was brought to the Quebec Conference by Col. Frederick S. Strong, Jr., from the China-Burma-India Theater. It was a map showing the planned line of communications from Calcutta in India to Kunming in China when the Burma Road was reopened. The width of the line indicated the tonnage per month passing any point. Different colors indicated the tonnages being moved by different means of transport. The chart showed .6 million tons a month going into Calcutta, but less than .2 million reaching Kunming. The .4 million tons a month that was diverted provided whatever was not available locally for the maintenance and operation of the line of communications, which included the port of Calcutta, the Bengal and Assam Railway, the barge lines up the Brahmaputra, the airlift across the Himalayas, and the roads and pipelines from Calcutta to Kunming. It also provided much of the support for the British forces covering India, for the three Chinese divisions being trained by the Americans, and for several bases for American long-range bombers. Although it would have been inaccurate to charge all tonnage diverted to these forces to the maintenance and operation of the line of communications, it was certainly reasonable to charge much of it because the protection of this line of communications was one of the principal reasons for these forces' being where they were.

In studying transportation in support of the invasion of
North Africa in World War II, we found that available cargo
shipping was inadequate to carry all the TO&E equipment of
the troops plus a reasonable increment of the theater reserve.
It was not, however, so inadequate as to make the operation
infeasible, and so I recommended that the theater agree that
the task force would go with only half its truck transport since
the initial American role was supposed to be static. General
Mark W. Clark (in Washington representing the theater commander) accepted this recommendation. When later developments required the U.S. II Corps to move into Tunisia, more
motor transport became a critical necessity. Because motor
transport was the only major shortage, we were able to expedite its shipment when shipping and escorts for an extra
convoy were found.

If the transportation system will support, or can be developed in time to support, the forces necessary to carry out the
operational plan, the rest of the logistics can usually be
brought into line within a reasonable time. This is because
Army units are designed to fight practically any kind of war
anywhere, and Army mobilization plans provide balanced
forces at the rate they can be moved overseas to Europe (the
nearest probable theater of operations). The expression
"within a reasonable time" is used because there have always
been peacetime shortages in materiel and logistic troops that
are usually considered and accepted as reasonable risks in the
development of the budget. Given the resources of the United
States, any such shortage can be eliminated in time, and this
time is always considered when accepting the risk as reasonable. Starting from scratch, almost any type of logistic unit can
be organized and trained in six months, and almost any item of
materiel can be produced in quantity in eighteen. These are
the maximum times. It is seldom that something has not been
done to help cut these times. A reserve logistic unit, for example, can usually be put in shape to go to a theater in three
months. A going production line can have its rate of production increased by being put on a two-shift basis. The production rate will not be doubled (an increase of about 75 percent
is usual) nor will the increased rate be effective immediately. A

gradual increase takes place over several months as personnel are recruited and trained and as subcontractors for parts and components increase their production.

The successful logistic support of contingency plans depends primarily on recognizing far enough in advance all requirements in the transportation field and any special requirements for service troops or materiel beyond those provided with a balanced force organized and equipped in the standard manner. Our first requirement, then, is to secure or produce such contingency plans far enough in advance to give time for organizing and training the necessary logistic troops and producing the necessary materiel. Although this time has varied in the past, depending on the requirements and availability of logistic troops and materiel and our capacity to organize and train troops and to produce materiel, we found in World War II that we could do very well if we had a year between planning and execution.

As Director of the Planning Division of the Army Service Forces, one of my functions was to keep a close liaison with the Operations Division of the War Department General Staff (OPD) and either secure contingency plans from them far enough in advance or prepare them in my own division. These contingency plans, when made, were given to the technical services, which then each computed its own personnel, troop, and materiel requirements to support the plans. Army Service Forces (ASF) headquarters required the technical services to develop a basis of assignment for each type of logistic support unit (such as so many units per corps, or one for so many items to be supported). Headquarters reviewed the proposed logistic troop requirements against the theater and overall combat troop bases and special requirements. It defended and presented the resulting logistic troop list to the Deputy Chief of Staff of the Army, who required justification for the units and reviewed the totals against the manpower resources available to the Army. ASF headquarters also reviewed the materiel requirements developed by the technical services and included them in the Army supply program. To the best of my memory, no·review by item was made by higher authority except by the War Production Board, which reviewed the requirements for raw materials and productive capacity with a view to reducing them to what was feasible for industry. Under wartime pressures, Congress had neither the time nor the staff to make an overall detailed review.

Although not prescribed as such in World War II, there was an effective overall limitation on the size of both logistic troop and materiel requirements. This limiting general factor was the availability of cargo shipping. In a global war with many theaters of operations, the basic allocation of strength to each theater was made by the annual allocation of cargo shipping to it. Logistic troops were furnished in a fairly definite proportion to combat troops, to combat support troops, and to air forces.

The original BOLERO troop basis of one million men to be transported to the United Kingdom for the cross-channel operation was made up under then Col. John E. Hull. I worked with the technical services to produce the logistic troop component. The troop basis came out in four approximately equal segments: troops in divisions, combat support troops, Air Corps troops, and logistic troops. This BOLERO troop basis was a model, the composition and proportions of which held up pretty well throughout the war. The Air Corps percentage came down a little, but I do not remember a distinct trend in any of the other segments. Troops requested by a theater were not sent unless they could be supported with the cargo tonnage allocated. The cargo tonnage available thus roughly set the ceiling, and within that ceiling we produced balanced forces and balanced logistic support.

To be satisfactory, a contingency plan does not need to be detailed nor even very accurate. We are not trying to outguess the enemy or do the theater commander's thinking for him; we are trying to have available the resources the theater needs. We found in World War II that if we had available the planned logistic resources for an operation, there were relatively few unforeseen requirements regardless of how the commander decided to accomplish the mission. We could expedite the provision for these unforeseen requirements. The fewer the unforeseen requirements, the more they could be expedited.

One of the early efforts to provide logistic troops and special logistic equipment to a theater in World War II was for the operations in the Gilbert and Marshall Islands. Long before any specific objectives in these islands had been selected, I and Captain Warlick, acting for the Navy, together drew up a logistic troop basis and a list of special equipment believed to be required. Lacking any better information, we selected a typical atoll from the *Pacific Pilot*, published by the U.S. Coast and Geodetic Survey, and studied it to provide a basis on which to estimate requirements. We agreed which service should provide each unit or item. We sent the list to the commander in the Pacific, who endorsed it and asked for its early delivery. Early delivery was accomplished because planning and action had started far enough in advance of the landings. The recommendations included some 20,000 logistic troops and such special equipment as amphibious tractors (to get across the coral reefs), dumb barges (to provide depot storage for supplies, which could then be moved from atoll to atoll without being unloaded ashore and then reloaded), and saltwater stills (there was little or no fresh water on any atoll). Captain Warlick accompanied the operations against Tarawa and Kwajalein and saw the effectiveness of the advance planning effort firsthand.

The War Department came out of World War II with a formalized planning system based on the Strategic Logistics Study. This document started with a scenario describing the campaign to be carried out and a combat troop basis. It was then supposed to develop requirements for logistic troops and special logistic equipment far enough in advance to permit the organization and training of troops and the procurement of special equipment.

The Strategic Logistics Study procedure has had its ups and downs since World War II. Three major influences hurt it. Tacticians are reluctant to forecast combat operations far enough in advance to provide for procurement items with long lead times. Logisticians themselves are inclined to make the whole procedure prohibitively laborious by going into a degree of detail that is unjustified given the inaccuracies that always accompany long-range tactical estimates. Budget specialists oppose procurement of items for which proof of need is tenuous.

During the stress of World War II, the Army Service Forces could and did develop contingency plans, determine the logistic requirements to support these plans, and then procure the necessary materiel and organize the necessary service troops. Such initiative, however, is not practicable in peacetime or even during limited wars such as Korea and Vietnam when funds and personnel spaces are closely regulated by the Department of Defense, the Bureau of the Budget, the Congress, and the President. I have long felt that there is a basic need in peacetime to develop contingency plans and procedures for computing logistic support requirements for those plans that are approved by or are acceptable to all these superior echelons as a basis for appropriation requests. As Deputy Chief of Staff for Logistics, I was unable to accomplish this, but I hope that a general recognition of its importance will come in the future.

I know of no contingency plans or any supporting strategic logistics study prepared in advance for the Korean War. The war was started with a surprise attack by an underrated enemy. The time from conception to execution was so short that we could only provide troops and equipment already available. Logistic troops to support the combat forces were sent from Japan, but they had to leave behind a considerable proportion of their strength to operate the logistic base remaining in Japan. Equipment and supplies had to come from what was left over from World War II. Fortunately this equipment included the landing craft needed for the Inch'on landing. Most of the equipment and supplies, particularly the motor transport, required major rehabilitation. Fortunately, again, this rehabilitation had been started two years before the war broke out, and although little in the way of serviceable reserves was on hand, the rebuild program was in operation and could be expanded.

The fact that Korea was fought without the prior provision of materiel and logistic troops does not prove that contingency planning is unnecessary. It only indicates that for a small war against an unsophisticated enemy the United States can provide most of the logistic requirements from such resources as it has kept available against the possibility of a major war. Even so, such a small war will likely be hampered and delayed a little by an inadequate number of service units in the Active Army and by a lack of special items of materiel for which requirements could have been foreseen and provided. For Korea, such deficiencies were corrected as expeditiously as practicable. Reserve logistic units, although inadequately trained, were called to active duty and sent to the theater. Airlift had to be used within Korea even for such heavy cargo as ammunition until the Korean National Railway could be put back into operation. This was delayed by a lack of railway troops as well as by a shortage of railway equipment that had to be produced in Japan, which had built the railway when Korea had belonged to it. Problems in the production of steel shell cases for artillery were overcome before the supply of brass shell cases, used and reused, gave out.

Unlike the war in Korea, the war in Vietnam came as no surprise. Our commitment, once initiated, expanded gradually. But so far as I know, there again was no approved operational plan with a supporting strategic logistics study to serve as a basis for procurement. Rather, requirements were developed in the theater and met as soon as practicable. Fortunately, there were no particularly critical needs for materiel not available from reserves. A need for helicopters and their supporting equipment did develop, but their production was already being expanded. On the other hand, the logistic troop situation was critical. The usual shortage of logistic troops in the Active Army troop basis existed at the beginning of the expansion. The President's decision not to call up Reserve units caused hastily organized, inadequately trained logistic units to be sent. Apparently, the theater logistic system is only now (late 1969) recovering from the effects of this decision.

Reasons why the problem of providing logistic support for contingency plans recur.

The provision of logistic support for a contingency plan is a complicated procedure. When any important part of the plan is not satisfactorily justified to those engaged in budgetary review, then all the needed logistic resources are not likely to be provided. When the logistic support being made ready in peacetime is inadequate, responsible logisticians will have to press for improved procedures that are logical, defensible, and understandable.

Form in which the problem of providing logistic support for contingency plans will probably recur.

The problem may recur in the form of a requirement to provide a contingency plan or scenario that is defensible as a basis for securing appropriations in peacetime for the provision of logistic support. It may recur in the form of a requirement for the improvement of any of the procedures by which the logistic troop and materiel requirements to support a contingency plan are determined. The problem may even recur in a somewhat similar form as a requirement to play a war game over and over with the same initial inputs but with varying tactics to develop the best use of a new weapon and to determine its capabilities (as was done recently in replaying a war game testing the tactics of the Cheyenne helicopter). The ultimate objective is to determine a contingency plan that will be reasonable if the new weapon is provided. In my opinion good contingency plans can best be developed by expanding a war game to include more logistics, by modifying it to make the tactical operations reflect the effects of any inadequacies in logistic support, and by slowing it down to a more likely rate of operations, rather than playing the same game over and over. If such a game is based on a reasonable and probable set of political and military circumstances, the line of action developed ought to make a defensible plan.

The problem could arise for any specific procedure in regard to the determination of requirements for items coming under the Army's procurement of equipment and maintenance appropriation. It could also arise for any group of procedures in regard to the determination of broad theater requirements or for any specific item in the support of a contingency plan. All of these possibilities seem to me to require competence in conducting war games that take into account the broadest range of possibilities.

CHAPTER 4

Career Management of Logistic Personnel

Reorganizations are generally directed toward solving certain specific problems. While doing so, they usually create new problems. Several years ago an Army reorganization reduced the functions of the Deputy Chief of Staff for Logistics and eliminated much of the Army's technical service system. Whether or not this reorganization solved whatever problems it was supposed to solve, it created some new problems, or rather reopened some old ones, by eliminating the solution that was in effect. I consider the most critical of these problems to be the career management of logistic personnel. This is also the problem to which I devoted the most attention when I was Deputy Chief of Staff for Logistics from 1955 to 1959.

It is axiomatic that the success of any organization depends heavily upon the selection, training, and utilization of its personnel. Until shortly before World War II, each chief of a combat army or a technical service, subject to rather broad guidance from the General Staff, exercised a strong influence on personnel selection and assignment through his personnel office and a strong influence on doctrine and training through his board, his school, and his technical inspections. For the technical services, the matter of selection and recruitment of competent officers was particularly critical. Graduates of the U.S. Military Academy were commissioned only in the combat arms, the Engineers, and the Signal Corps. The Ordnance Corps, the Quartermaster Corps, the Chemical Corps, and, later, the Transportation Corps had to depend for officers on ROTC and on transfers from the combat arms. Each of the Chiefs of Technical Services could influence the choice of a competent officer by his personal interest. The evidence of personal interest often took the form of an indication by the chief that the officer concerned would be given certain future assignments. Advanced study in a civilian institution was one of those assignments most sought. The prestige of the service chief was also a factor, given the power he wielded and the fact that he had been selected as the outstanding man in his specialty. As a young lieutenant of Field Artillery, for example, I thought of the Chief of Field Artillery as being at the top of my profession. I did not even know the names of the heads of the General Staff divisions. I recognized the Chief of Staff, General John J. Pershing, as a great soldier, but he, like the rest of his staff, were too remote to be of other than historical interest to me or most other junior officers.

After the elimination of the Chiefs of Combat Arms before World War II, the difference between the handling of officers in the combat arms and the technical services became quite marked. The difference was basically in the degree of personal attention that the officer at the top of his profession gave to the selection, training, and assignments of the officers over whom he had supervision. Throughout World War II and Korea, G–1 sought to perform these tasks for the combat arms largely based on career records—instead, as before, on personal knowledge. Perhaps I can best express the difference this way: if a major unit commander needed an Ordnance officer, he could be confident that the man offered to him had been carefully selected, probably by the Chief of Ordnance himself, as having the training, background, and characteristics that would make him a good Ordnance officer for the unit. If the commander needed an operations officer, he could be confident only that the man offered to him by G–1, War Department General Staff, was from one of the combat arms, was of the proper grade, and had probably had some experience or training that might fit him for the assignment.

The combat arms officer assigned then had to sink or swim, but commanders were not enthusiastic about trying out individuals in key positions who might or might not be able to perform adequately. If they proved inadequate, it might have a serious effect on the performance of the unit. As a result, a practice developed under which commanders made individual name requests, based on their own knowledge or that of their staff officers. While more satisfactory to the commander with a vacancy to fill, this asking for individuals by name had the disadvantage that the individual usually was already serving under some other commander, who was likely to feel that his own mission precluded the release of the officer. During World War II and Korea no one acting weighed the importance of the two positions for the good of the service as the Office of the Chief of Field Artillery would have done. Where the considerations supporting a transfer were particularly strong, an appeal by the requesting commander or by the individual himself to his next higher superior, thereby reaching a more disinterested level, was sometimes used, but this was accomplishing an action in spite of the system rather than because of it.

With the elimination of most of the Chiefs of Technical Services, the personnel of most of those services are now (mid-1969) in a similar but even less desirable position than the personnel of the combat arms have been in for many years. They are specialists, and in an age of great technological progress, specialization is essential. Specialists are best judged and guided by men successful in their specialty. Nevertheless, decisions on individuals are no longer made by a respected service chief or his service personnel officer who knows many individuals either personally or by long familiarity with their records and whose success in office depends largely on how well these individuals are used. No personnel specialist, however competent, has prestige approaching that of the former Chiefs of Technical Services. No computer can make the people it manages feel that it has a personal interest in or understanding of their careers.

With respect to the logistician who is a generalist, the man who is to become a G–4 at any level of command, the situation is particularly bad. So far as I know, before I became Deputy Chief of Staff for Logistics no one had ever tried to guide the careers of individuals in the generalist logistic field. As a result, generalist logisticians were being developed only by chance. It was certainly so in my own case. After eighteen

years of service, none of which was in logistics, I was assigned
to the G–4 Division of the War Department General Staff in
June 1941. Thereafter I spent most of my next twenty years of
service in logistics because I had gained experience in the field
while few others had. When I became Deputy Chief of Staff for
Logistics, I sought to manage the development of generalist
logisticians.

My efforts at career management in the generalist logistic
field recognized the preeminence of combat arm or technical
service. The technical services had good career guidance for
developing officers in their own specialty. The Deputy Chief of
Staff for Personnel was endeavoring to establish an adequate
system for the combat arms. I accepted that officers chose
their arm or service because they desired assignments with the
troops, the schools, or the agencies of the chosen arm or
service. I found, however, that there were so many branch-
immaterial assignments in the Army that on the average an
officer spent about half his service after reaching the grade of
major in such assignments. Logistics could then be a secondary
Military Occupational Specialty (MOS). I therefore sought to
avoid disturbing officers' assignments with their own arms or
services. Rather, I sought to interest officers in the grade of
major or above who had performed well in one assignment in
logistics in having assignments in logistics alternate with assignments in their own arm or service. I undertook that as
long as they remained in my logistics career program, I would
endeavor to assure them that their assignments in logistics
would become more and more challenging. In addition, I undertook to add my influence to that of their arm or service in
securing for them whatever school assignments they were
ready for and whatever promotions they had earned. To assist
me in this effort, I asked to be assigned to all the general
officer promotion boards on which I could legally sit, which
was half of them. This was a heavy and time-consuming task,
but studying and restudying the records of hundreds of colo-
nels and generals gave me a good knowledge of all the senior
officers with logistics experience and enabled me to give support to those who were deserving.

In order to be able to arrange for assignments of ever-
increasing responsibility, I had to convince all the senior com-

manders in the Army that the best way for each of them to obtain a competent G–4 staff officer or logistic commander was to take one of the officers I nominated. I handled the senior logisticians personally. When I left the Office of the Deputy Chief of Staff for Logistics, every important logistic position in the Army was held by a logistician nominated by me and accepted by his commander or, in one case, suggested by his commander and concurred in by me.

With respect to technical service officers who preferred to remain strongly service oriented rather than join the logistics career program, I supported the service chief in his management of their careers by helping to secure desirable school assignments and promotions for those who were worthy. I went beyond this only in exceptional cases, such as those of officers who might become candidates themselves for chief of a service. Because I was Deputy for Logistics, I was also president of the board that nominated officers to become one of the Chiefs of Technical Services. I therefore sought to know personally and be familiar with the records of the brightest prospects, and I sought to see to it that these officers were given assignments that broadened their backgrounds to fit them better for the top positions. No officer nominated for chief of a service by one of these boards during my tenure ever failed of selection.

I finished my assignment as Deputy Chief of Staff for Logistics by recommending to the Chief of Staff and the Secretary three officers well qualified to take my place: a former technical service chief, whose assignment as a deputy Army commander I had arranged in order to broaden his background; a logistician with a broad background, whose assignment as an overseas communication zone commander I had arranged to give him some high-level command experience as well as more logistical experience, and my deputy, whose previous assignment as commander of an overseas communications zone I had arranged to give him command experience and confidence. One was selected to replace me. The other two became Army commanders, proof that their work in logistics had not handicapped them with respect to assignment in other fields. I do not present these instances to show myself in a favorable light. The various officers concerned had shown themselves outstanding before I ever exerted any influence. I only sought to ensure that they were prepared for higher assignments in logistics. The Deputy Chief of Staff for Logistics, however, no longer has the authority over logistic officers personnel selection that I had and, therefore, not the same opportunity to exert personal influence.

My efforts at career management for Department of the Army civilians in logistics were similar to those for officers but more restricted. While transferring civilians to desirable assignments in the Office of the Deputy Chief of Staff for Logistics from the field or from the offices of the Chiefs of Technical Services was practicable; the reverse was not because of the lower grade structure. As a result, desirable assignments outside the Office of the Deputy Chief of Staff for Logistics had to be sought in the offices of the Army secretariat or the Department of Defense. One problem that I sensed with respect to civilians in my own office was that even the best of them were not accorded respect as readily as were their military counterparts, and this limited their effectiveness. In order to correct this, I sought to build the prestige and broaden the outlook of a few of the truly outstanding civilians by sending them, one at a time, to the Industrial College of the Armed Forces or the National War College. This was a painful process insofar as current workload because no replacements were allowed. However, the objective was attained. The performance and increased status after graduation of those who were sent to the highest military schools has been gratifying.

With respect to technical service civilians, assignments overseas had caused a continuing problem. When such civilians returned from overseas, there was not necessarily an appropriate assignment in grade available for them. There were two exceptions. Ordnance had developed a career field for ammunition specialists. Personnel in this specialty were assured appropriate assignments whenever they returned to the United States from overseas. The Corps of Engineers, where an engineer district was established overseas under the supervision of the Chief of Engineers, also assured its civilian personnel appropriate assignments on return to the United States.

Civilians were predominant in many technical fields in which their qualifications were needed overseas, yet these civilians were understandably reluctant to accept overseas assignments with no assurance of acceptable assignments on return to the United States. When they did accept assignments overseas, they often became virtual exiles and stayed overseas almost indefinitely. I sought to have all the technical services arrange assignment methods that would accomplish for all their civilians what Ordnance had done for its ammunition specialists. I had a civilian personnel directorate within my own office to help me with this. The Deputy Chief of Staff for Logistics no longer has such a directorate.

My efforts at career management for technical service enlisted men included requiring the offices of each of the Chiefs of Technical Services to establish a career pattern in each specialty in their service that would permit a really good man in any specialty to advance in his own or a related specialty to the top enlisted grades. The Deputy Chief of Staff for Logistics no longer has the Chiefs of Technical Services to work through.

Reasons why the problem of career management of logistic personnel recurs.

In a technical age the Army needs specialists. It appears to me that the lack of personal attention of individuals in high positions to the careers of technical personnel will cause the Army to lose some of its best specialists from dissatisfaction and will reduce the ability of the Army to develop new ones of equal quality. Poor logistic performance will indicate the need to secure and train such specialists.

Form in which the problem of career management of logistic personnel will probably recur.

Just as shortages of doctors have in the past prompted the Army to study how to secure, train, and retain doctors, so will the shortage of qualified logistic personnel prompt the Army to study how and take necessary action to secure, train, and retain them.

CHAPTER 5

Operation of the Logistic System

There are many problem areas in the operations of any system as complex and as dispersed as the logistic system. I have selected for discussion in this chapter some of the more important problems not covered elsewhere.

Coordination of Command and Technical Authorities

Before the reorganization that eliminated some of the Chiefs of Technical Services, the Army had two concurrent lines of authority. In the so-called chain of command, commanders issued orders to next subordinate commanders to control operations. There were virtually no restraints on this chain of command. In the so-called technical channels the staff officer of a technical service issued technical guidance related to the functions of his service to the staff officers of his service at the next lower headquarters and to units of his service reporting to his headquarters. Technical information flowed back up this same channel. Thus it was proper for a commander in the chain of command to direct an Ordnance maintenance unit to give first priority to the repair of medium tanks. It was equally proper for the Ordnance officer on the staff of that commander to give guidance to the unit on when and how to replace a tank turret.

In this system, the Chief of Ordnance commanded a board and various proving grounds that made technical studies and tests, arsenals that did research and development and a little production, and schools that helped develop and teach Ordnance technique. He had technical inspections made of the operation of Ordnance units and of the care and use of Ordnance equipment in the hands of other units. Through his technical channels he issued guidance on Ordnance matters and received through the same channels information on how Ordnance equipment was functioning under varying field conditions. He also received information on enemy ordnance. This flow of technical guidance down and information up was generally understood throughout the service and operated without conflict with the chain of command. It gave a technical service chief information upon which to base technical guidance, research and development, and retrofit decisions. The elimination of many of the chiefs of service and the introduction of functional units left the technical channels inoperable.

Reasons why the problem of coordination of command and technical authorities recurs.

Without the flow of information and guidance down and information and suggestions up through technical channels, research and development, retrofit, repair, and operation of equipment will suffer.

Form in which the problem of coordination of command and technical authorities will probably recur.

The Army will need to seek a standard method of providing technical guidance from the Commodity Commands to the field and of securing from the field information for the Commodity Commands on technical performance and suggestions for improvement without conflicting with or overburdening command channels.

Monitoring the Operation of Overseas Supply

One of the major steps forward taken in the effective supply of overseas theaters in World War II and continued in the Korean War was the establishment of an Overseas Supply Division at each port of embarkation. Early in World War II some theater commanders were allowed to have "rear echelons" either in the War Department or at the port of embarkation supporting the theater. These rear echelons lacked authority and sometimes caused confusion because they were concerned with supporting only their own theaters. They were replaced by the Overseas Supply Divisions, which had both authority and expanded functions. To correct errors in requisitioning, an Overseas Supply Division edited theater requisitions against consumption rates already experienced, modified to provide for projected operations; to expedite shipments, it extracted requirements and forwarded them to appropriate depots, received reports when shipments were ready, and followed up on delays; to avoid jamming the port, it called forward shipments according to desired priorities of delivery and ships available; to ensure that ships were loaded "full and down" (cargo space filled and ship loaded down to the Plimsoll marks) and that high-priority cargo was accessible for early unloading, it participated in planning the loading of ships; to facilitate theater supply operations, it forwarded cargo documentation well ahead of a ship's arrival overseas; and to ameliorate the effect of a ship's sinking, it promptly ordered forward replacement cargoes.

The Overseas Supply Divisions operated under the authority of the War Department. Thus, many War Department responsibilities for theater resupply were concentrated in one agency to which a theater commander could look to fill his needs. The Overseas Supply Divisions maintained close liaison with the theaters they were supporting so as to be familiar with theater problems and also be prepared to provide unusual or exceptionally large requirements for projected operations.

Reasons why the problem of monitoring the operation of overseas supply recurs.

The Overseas Supply Divisions were discontinued about 1962. Many difficulties in the supply of Vietnam and other overseas commands have occurred since in the areas in which the Overseas Supply Divisions had operated. Ships have arrived in overseas ports in greater numbers than could be unloaded promptly. Undetected errors in requisitions have caused unwanted deliveries. Lack of follow-up has caused failures and delays in delivery. Unmarked shipments have arrived. Documentation has often failed to arrive in advance of ships and has often been inadequate. Many other difficulties of the types the Overseas Supply Divisions were organized to eliminate would probably have occurred had Vietnam been a major war.

Form in which the problem of monitoring the operation of overseas supply will probably recur.

The problem will probably recur as a requirement for a means to correct one or all of the deficiencies mentioned above or as a proposal to fix "throughput" responsibility in some agency, which would accept supplies from the supplier or depot and deliver them to the military consumer overseas.

Application of Funds For Operations and Maintenance

In peace, field commanders finance those operations that require local financing with their allocation of funds from the Appropriation Operations and Maintenance, Army (O&MA). Such funds are seldom available in adequate amounts, and so determining the priority of their application is required and usually presents many difficulties. This problem does not arise overseas in war. In war, overseas field commanders finance local procurement of goods and services by a great variety of means, usually determined by expediency but basically without restriction.

Among the many purposes for which O&MA funds may be expended, deferred maintenance of real estate, minor construction, and purchase of repair parts seem to me to have given the most trouble. These purposes compete with each other for the available O&MA funds.

Deferred maintenance of real estate is one purpose for which the requirements always seem to exceed the available funds. Deferred maintenance is reported to the Congress in each annual budget presentation as part of the defense of the O&MA appropriation request. At various times the House Appropriations Committee has insisted that enough of the operations and maintenance appropriation be used for deferred maintenance to reduce the backlog, but such a reduction has seldom been accomplished. One reason is that the Army retains many structures long beyond their economic lifetime because replacement through the appropriation Military Construction, Army (MCA), is so difficult. The second reason that the backlog of deferred maintenance increases is because maintenance of real estate has relatively low priority compared with the other purposes for which locally available O&MA funds can be used. A third reason is that there are no very clear or accurate standards by which to repair or renovate and no way of estimating accurately the cost of repairing structural parts that are inaccessible until repair is started. As a result, preliminary estimates are usually low.

Minor construction is also a purpose for which requirements always seem to exceed the available funds. This is partially because remodeling and extension, which are the most common types of projects under minor construction, are always difficult to justify. It is partially because a construction project using O&MA funds can be started at once, whereas it takes about three years to get started if funds are sought through the MCA appropriation. Some control in minor construction has been attempted by limiting, for a single project, amounts that are within the authority of various decision levels. This restriction on size without influence on type and number of construction projects has not proved to be a very valuable control.

The procurement of repair parts is a third purpose for which the requirements always seem to exceed the availability of funds. I have never seen a reasonable justification for restricting the use of repair parts: failure to replace an unserviceable part reduces the efficiency of a much more expensive mechanism and risks damage to other parts. Nevertheless, such a restriction exists because of the requirement that local commanders finance their requisitions for repair parts from their allocations of O&MA funds. This has proved to be a bad type of control.

These three purposes discussed above, then, are purposes for which requirements usually exceed available funds. They are also purposes that compete with each other for the inadequate funds available. Priority among them is a matter of judgment for the responsible commander. In Korea, I inherited a theater reserve that contained a great deal of unserviceable Ordnance equipment left unrepaired for lack of parts. I also inherited a strong commitment to minor construction. The former commander had placed a lower priority on repairing unserviceable equipment to achieve logistical readiness for combat than on constructing facilities to improve the troops' standard of living so that their morale would be higher and they would be more likely to reenlist. Maintenance of facilities habitually occupied a low priority since it contributed less directly to military effectiveness.

Reasons why the problem of the application of O&MA funds recurs.
In the past there has always been a shortage of O&MA funds, so the problem of how best to apply them continually recurs. It is generally accepted that authority for deciding their application ought to be largely decentralized. But it is possible that a limited degree of centralizing influence could be exerted to good purpose by establishing some standards in each of three areas discussed.

Form in which the problem of application of O&MA funds will probably recur.
The form in which the problem will probably recur is in a requirement to determine what influence should be exerted by central authority on the application of O&MA funds and how it should be exerted. More specifically, the requirement may call for determining what standards can and should be set by the Department of Defense and the military departments to influence how much is done toward each of the three purposes discussed.

Use of Stock Funds for Repair Parts

The original proposal for establishing stock funds provided that they would be revolving funds. Capitalization would come from the stock fund's assuming ownership of repair parts already held by the Army and already paid for by the Army. Army agencies would buy repair parts from the stock fund with new O&MA money. These funds would then be available to buy new repair parts at the Army's discretion without restriction.

The stock funds have never operated in the manner proposed. Most of the money accumulated in the stock fund was withdrawn, as was the Army's discretionary authority to expend what was left. These actions eliminated all logistic advantages from the stock fund and left one major logistic disadvantages, the stock fund in the communications zone of the European Command. This was an experimental downward extension of the stock fund that, in my time, I had been unable to block. It is still operating in 1969. Under it, the Seventh Army in Europe must buy repair parts from the stock fund that owns the repair parts in the communications zone, paying for them with the O&MA funds allocated to Seventh Army. When funds run low, review of requisitions has to be resorted to in order to devote funds to the most important requirements. This causes delay in repairing even high-priority items and causes turbulence in the system as parts already bought are turned back in for credit with which to finance the purchase of other parts. When funds run out, equipment goes unrepaired even though the required parts are in the theater.

Reasons why the problem of the use of stock funds for repair parts recurs.

Any effort to improve the operation of the supply system in filling requirements for repair parts will find the overseas stock fund in Europe to be a handicap and will probably find the stock funds in the United States to be of questionable value.

Form in which the problem of the use of stock funds for repair parts will probably recur.

The problem of stock funds will probably recur either in connection with studies to improve the effectiveness of repair parts supply in Europe or in connection with any study that examines whether the tremendous amount of bookkeeping and reporting that the Army now does produces results commensurate with the diversion it causes of supervisory attention and manpower from training and support of combat.

Handling of Excess and Surplus Property

No one with any knowledge of logistics thinks that the building up of excesses can be avoided. When Lt. Gen. Wilhelm D. Styer, Chief of Staff of Army Service Forces in World War II, was asked how excesses could be avoided, he is reported to have answered that he did not know but one thing he did know was that the side that won this war was going to end up with great excesses and the side that came out without excesses was going to be the side that had lost. General Styer did not mean that reasonable efforts should not be made to avoid excesses—he was merely recognizing the facts of life, some of which are covered below.

To the military in a combat theater an item in excess is of little consequence. If movement is necessary, excesses can be left behind. To the military at department level excesses are of consequence only if resources used in their production caused shortages of some other needed item. On the other hand, a shortage or even a threatened shortage can have a critical effect on a campaign. A prudent military commander is always looking over his shoulder to see if adequate resupply is coming in. If resupply appears questionable, he will probably institute rationing, which reduces combat effectiveness, or he will slow down operations, or both.

It is therefore to be expected that excesses will accumulate in an overseas theater during a campaign. These can, however, be limited. One measure toward this end is the requirement for adequate inventorying that should identify accumulated excesses so that they can be drawn down. A second measure is the requirement for a reasonably accurate reporting of the expenditure of materiel by cause or purpose to support the determination of replacement factors. Good replacement factors improve the accuracy both of theater requisitions and of plans for the procurement of new materiel in the United States. A third measure is the requirement for an editing agency removed from the pressure and confusion of active operations. This agency helps eliminate gross errors that not only handicap theater supply but also introduce corresponding errors in the requirements for new production.

It is important to distinguish between, on one hand, excesses that may develop in an active theater because supply is greater than consumption or in the United States because inaccurate replacement factors provide the basis for procurement plans and, on the other hand, surpluses at the end of a war. The first can be limited. The second cannot be avoided.

At the end of the World War I, the U.S. Liquidation Commission, operating under the War Department but not under the Commander in Chief, American Expeditionary Forces (AEF), promptly sold all the property left behind by the AEF to France for a lump sum. Although most of this equipment had been heavily used and much was becoming obsolete, as does most equipment at the end of a war, various congressmen were quit critical of the sale. Because of this congressional criticism, Lt. Gen. Brehon B. Somervell at the end of World War II asked that a civilian agency dispose of surplus property overseas. The Foreign Liquidation Commission was created to perform this function under the Department of State.

After the defeat of Germany, the best equipment of our forces in Europe was shipped to the Pacific, where the war continued. The forces expected to remain in the occupation of Germany were equipped with the best materiel remaining. Certain items, mostly weapons, were returned to the United States for war reserves. In the European Theater, a force of some three million men—which had gone overseas with about two short tons per man of initial equipment and had built up behind it perhaps another ton per man of similar materiel— shrank rapidly as political pressure to bring the troops home grew in the United States. The occupation force was finally stabilized at about 100,000. Troops returning home turned in all of their equipment except some of their personal equipment. This left me, as G–4 European Theater, with something like six million tons of excess and surplus property spread over France, Belgium, and Germany. That in France and Belgium was in the custody of the Western Base, which had a strength of some twenty thousand men. As the redeployment of the main forces approached completion, it would be possible to redeploy the troops of Western Base also, provided the property in their custody could be disposed of. It was quite apparent that the cost of paying and maintaining the twenty thousand men and renting the depots they were occupying would soon vastly exceed any return we might obtain from the sale of the excess and surplus property. I therefore had the theater chiefs of technical services ship to our depots in Germany whatever equipment and supplies still remaining in France and Belgium they thought the occupying forces might use. Admittedly, this flooded our depots in Germany, but it also left our depots in France and Belgium ready to dispose of what was now only surplus on hand.

The Army wanted to have the surplus property disposed of as rapidly as possible. The Foreign Liquidation Commission quite understandably wanted to get the best price possible. In spite of this difference in objectives, the two agencies cooperated well. I have only praise for the Foreign Liquidation Commission. It did its work well, yet I still question whether a quick bulk sale to each country in which we had surplus property, as was done in World War I, would not have been more economical. The savings in personnel costs and rentals would have been considerable. On the other hand, returns from intergovernmental sales in both wars were probably never collected, as most war debts were forgiven or allowed to fade into oblivion.

The equipment and supplies shipped to Germany from Western Base were packed hurriedly by foreign employees or American personnel eager to go home. Packing was poor and identification was often lacking. The same was true of equipment and supplies turned in by our redeploying troops to our depots in Germany. The problem then became one of identification and of evaluation of condition. Most of the problem was ordnance; we had a major operation in unpacking, cleaning, identifying, and evaluating these items at the Mannheim Depot in Germany. Again, as in the case of the retail sales of surplus property, I cannot prove that this operation was uneconomical. I just think it was. Both were highly successful politically. There was relatively little congressional criticism and even some praise for our handling of surplus property.

At the end of the Korean War the disposal of equipment left behind by our redeploying troops offered little difficulty because we used it to complete the equipment of the Korean forces and to replace such of their equipment as was not economical to repair. Most of the equipment reserves for our forces in Korea, including much rebuilt equipment, were held in Japan during the Korean War and used thereafter for Lend-Lease throughout the Asian fringe of the Communist bloc. I did, in Korea in 1959, inherit a couple of large motor truck graveyards. We had furnished equipment to the Koreans with a proviso that when no longer required it should be returned to us. Therefore, worn-out trucks, stripped of all usable components, accumulated. Unable to sell them, we were only able to get a steel mill to take them off our hands if we helped haul them to the mill.

Before we intervened in Vietnam with major forces, the French pulled out leaving large quantities of unserviceable equipment, again returned to us under a Lend-Lease type of agreement. The government official who first saw these dumps called them "acres of diamonds," and soon the Army was shipping these "diamonds" to Japan for rebuild. Once more I have no proof that this operation was uneconomical, although I think it was. It was successful politically.

Reasons why the problem of handling of excess and surplus property recur.

There have accumulated in Vietnam quantities of unidentified supplies and of unserviceable equipment. As I write this in mid-1969, some of this property is being outshipped to Okinawa for identification or to Japan for repair, but, as our troops redeploy, more property will accumulate. How to handle this property economically and without political repercussions is a problem that is already with us and one that will probably become more acute.

Form in which the problem of handling excess and surplus property will probably recur.

The problem will probably recur as a requirement to establish guidelines on how excess and surplus property in Vietnam, Okinawa, and Japan shall be segregated for identification, repair, or disposal and on how disposal shall be accomplished.

Management

We in the military have studied and practiced management all our lives. The exercise of command is management. Unfortunately, when management recently became a fetish, all managers seemed to learn the virtues of centralization, computerization, use of all-powerful project officers, and cost-effectiveness, but not to understand their limitations. These are procedures that are useful and easily applied in a simple and routine business. But the problems of war are neither simple nor routine. For the military, I feel it is usually better to decentralize, often better to approximate, almost always better to restrict the authority of project officers, and never permissible to measure cost only in dollars.

Among military matters, logistics is particularly complex. Decisions should be made at those points where there is understanding, and only on the broadest logistic subjects is there understanding at a high level. One of the best examples of good management that I have ever seen illustrates this point When I was G–4 of the Mediterranean Theater of Operations in World War II, Maj. Gen. Otto Nelson came from Washington to be deputy theater commander. He brought with him a group of "efficiency experts." They questioned everything. They sought suggestions, from the lowest levels as to how operations should be conducted. They had their own suggestions too, although none of them had ever seen a theater of operations before. I remember that, in that period, time and motion studies had shown that a man surrounded by a semicircle of filing cabinets that he could reach without leaving his chair was the height of efficiency. This arrangement was advocated with the same fervor that is now accorded the computer. Since I had no power to stop this harassment, I didn't try. Instead, I prepared to present my cause to the theater commander. I never had to. As everyone's patience was nearing exhaustion, General Nelson sent the efficiency experts home. I awaited a detailed directive. None ever came. Not even a single suggestion. General Nelson had just made us take a careful look at our own operations and left corrective action up to the judgment of the operators, the people who knew what they were doing. They were thoroughly stimulated. They took, of their own initiative, many desirable actions that they would have resisted had the actions been directed. They also prevented a number of undesirable actions that would have been forced upon them had they been directed by centralized authority. This experience illustrates an approach that was remarkably effective because it avoided the disadvantages of excessive centralization, or, to use Secretary of Defense Charles

Wilson's terms, the disadvantages of "a concentration of ignorance."

With the inability of the generalist logistician to acquire a thorough knowledge of all the field under his jurisdiction, a logistic commander is at the mercy of his subordinates. It was my practice to try to keep the decision making down where the knowledge was, but in the development of broad policies and the handling of problems that impacted on several areas, I had to use my own staff. Since, in many cases, I could not learn enough myself to make the best decision, my practice was to severely cross-examine the staff officer who had presumably gone deeply into the problem, this in order to satisfy myself that he had done his homework. Such a procedure was necessary for me but unpleasant for the staff officer and brought me a reputation for being hard on my subordinates. Once a staff officer had established his reputation with me for accuracy, thoroughness, and good judgment, I abandoned this practice. Thereafter he not only had freedom from cross-examination, he also had my best efforts to foster his career.

When a war starts, many more decisions have to be made and made quickly than can be made by the top officials. If the lower echelons have not been used to making decisions in peace, they won't make them in war. For example, before World War II, we in G–4 War Department General Staff, had been required to include a paragraph in every staff study, stating how much the recommendation would cost. We had to obtain the initials of an officer in our Fiscal Section indicating the appropriation to be charged and certifying that the funds were available. With that background, I received quite a shock a day or two after Pearl Harbor when, as General Somervell's night executive, I saw him allocate to the corps area commanders some $300 million. This was an action that had been cleared within the War Department General Staff but to the best of my knowledge had not yet even been put in a supplementary budget, much less appropriated. This was completely contrary to all peacetime practice, yet it was a very necessary action. I doubt that anyone in G–4 except General Somervell would have even thought of such an action. Fortunately, he had operated with wide latitude as Works Projects Administrator for New York.

In the making of decisions in areas where his knowledge is inadequate, a logistician may require a good deal of data, and now this seems always to call for a computer. However, care must be taken that the computer is not misused. A computer is a wonderful instrument when used for actions of a repetitive nature with programs that have been thoroughly checked out. Nevertheless, although many decisions on important logistic problems can be made based on a data bank derived from the mass of routine reports that is characteristic of computerization, many cannot. This is not because the computer is not accurate, but because in war so many variables are introduced that a routine report may produce data not all on the same basis. Such data are not suitable for processing. A good example occurred when we were gathering data on ammunition expenditures by mission in Vietnam under the COLED–V (Combat Operations Loss and Expenditure Date–Vietnam) project. The reports had been showing artillery ammunition expenditures subdivided by the mission being supported, such as search and destroy, clear and hold, and security and base camp defense. Then reports started coming in with an added subdivision: harassing and interdiction. Since artillery expends ammunition in harassing and interdiction fires in support of all types of missions, this subdivision rendered the data being gathered unusable until new definitions were put into effect. Similarly, when the theater changed the nomenclature of offensive missions from two (search and destroy and clear and hold) to three (search and destroy, cordon and search, and clear and secure), no one had the information necessary to redivide the data gathered under the two old missions so that it would fit under the three new ones. As a result, we could not combine the old and new data into a data base. Any routine reporting system must be continually policed to insure that apples and oranges are not processed as if they were the same.

I would like to see the ammunition expenditure system, which has been developed by the Army and the Research and Analysis Corporation working together, continue throughout the war in Vietnam and thereafter in peace. This would ensure that at the beginning of the next war valid data in this critical area would be accumulated properly. Since this is not to be, I

ATION OF THE LOGISTIC SYSTEM 71

hope that the reporting system will be taught in the Army's schools. This is the only method I know by which a practice not in daily use can be kept alive in peace ready to be put into operation effectively in war without a period of policing, explanation, and interpretation.

A second concern in using routine reports to gather data bases for use in computers results from the new problems that continually arise in war. Material gathered by routine reports designed to produce data for use in the development of solutions to a certain set of problems is not necessarily usable in the development of answers to new problems. I have found that the most useful report is one called for to help in solving a specific problem. In initiating such a report, explanations must be offered and interpretations made in light of the intended use of the collected data. Such a report must be policed through several reporting periods, its provisions thoroughly clarified, and explanations from reporting agencies for indicated shortcomings considered, all with respect to the specific problem to be solved, before the report will serve its purpose well. A commander can act only on a relatively few problems. Reports not bearing on one of these problems—and therefore not likely to be called to the attention of the commander to whose headquarters they are submitted—do not deserve attention at lower levels in competition with all the other demands of war. Such reports become both inaccurate and inapplicable to the new problems brought about by continually changing circumstances. Situations change so rapidly in war that data being reported in successive months may not be on the same basis.

Finally, time is almost always critical. Many important decisions in war cannot await the accumulation of the exact data that a computer requires. Accordingly, when available data have not been gathered to help in the solution of the specific problem at hand, it is usually better to use a quick approximation and get the decision made promptly.

Along with centralization and computerization, I consider the management practice of using powerful project officers to be misused. Currently a project officer is a czar authorized to make far-reaching decisions that are bound to affect many projects other than his own in ways that he may well not understand. The commanders in whose fields the project officer operates may suffer critical interf:rence in many areas as a result. Great authority is proper only for a project officer handling a project of top priority such as the Manhattan project, which developed the atomic bomb, or perhaps the Polaris project, which developed the submarine-launched missile. Using project officers with broad authority for many projects at the same time, however, is a dangerous practice. Virtually no priority is ever absolute. Responsible commanders must therefore exercise judgment in the degree to which the requirements of one project are allowed to infringe on the requirements of others, even those of lesser priority.

The proper function of a project officer, in my opinion, is that of an expediter. He takes part in the development of reports and schedules. He has access to everyone but no authority to give orders. He detects problems through his schedules and reports and through frequent visits to all agencies and installations working on his project. He brings these problems to the responsible commanders, going as high as necessary to get action. He may suggest solutions but cannot dictate them. He follows up energetically on decisions, and he reports their effects and progress to the responsible commander.

No discussion of management can well omit the measurement of cost against effectiveness. This is a perfectly valid procedure, but it is subject to misuse with relation to the military. For commercial transactions, the cost factor is properly measured in dollars because the basic purpose of commercial transactions is to make a profit in dollars. If a firm cannot make a profit, it fails. For military transactions, the cost factor should be modified because the basic purpose of military transactions is success in war. Accordingly, effectiveness, in addition to having a relationship to dollars, also bears a relationship to lives lost, to lives blighted by wounds, and to the effects of a national defeat. Since the value of lives, health, and victory is difficult to determine, it is usually desirable to measure cost against effectiveness only to decide which of two roughly equally effective systems should be acquired and in other cases to provide the best system that can be developed but to do so at the lowest reasonable cost.

Reasons why the problems of management recur.

New and successful commercial management practices must always be considered for possible adoption by the military. And as such practices already adopted by or imposed on the military prove disadvantageous, the question will be raised as to the desirability of abandoning or changing them.

Form in which the problems of management will probably recur.

The problem will probably recur as a requirement to determine whether or not a new management procedure should be adopted by the military or whether the results being accomplished by an existing procedure are worth the loss in strength caused by the diversion of effort from matters more closely allied to war.

Integrity

Throughout the military services, integrity is essential to operational effectiveness. An officer making a decision must be able to rely on any information he receives from another in the military as being the truth, the whole truth, and nothing but the truth to the best of the individual's ability to observe and report. This degree of integrity is of course required of a logistician as well as of any other military officer, but beyond this, a logistician should so conduct himself as to avoid any possible implication of a conflict of interest.

Ever since the start of World War II, military appropriations have been large. Logisticians handle most of these funds, including those for procurement, for operations and maintenance, and for construction. The public and the Congress are extremely sensitive to the use of these funds because of their impact on local economies. In the obligation of these funds, judgment exerts a major influence. This combination of big money being applied in sensitive areas and being applied based in large measure on judgment is bound to produce many challenges by disappointed contractors and disappointed localities.

Logisticians are also involved in disposing of surplus property, in reducing or discontinuing going operations that carry large payrolls, in closing installations and in disposing of real estate. Here, too, are large transactions, in sensitive areas being heavily influenced by judgment, and here too are protests and complaints.

Our best defense is a reputation for adhering to an exceptionally high code of ethics. It is not enough for a logistician to be honest. It is necessary that he so conduct himself that there is no basis for even the suspicion that anything other than the best interests of the military service and of the nation have been allowed to influence any official transaction.

It may seem unfair that there should be a higher standard of integrity demanded of us than that prevailing in the business world or even than that required of the civilian officials of the government, but whereas lack of popular support may only cause one politician to be replaced by another, lack of popular support may cause the military to dwindle to a dangerously inadequate level of strength. We in logistics can retain the confidence of the public and of our superiors and the respect of our subordinates only if every transaction is completely above suspicion.

When I was Deputy for Logistics I took pride in the realization that no justifiable negative reflection on the integrity of any logistician ever came to my attention. On the other hand, there were actions taken to insure the Army against any possible embarrassment. For example, one of the Chiefs of Technical Services felt that it was not becoming to his office to disqualify himself when a decision was required that might influence one of the companies in which he held stock. On his own initiative, he therefore sold his stock in any company that had dealings with the Army. His wife, who had inherited some such stocks many years before, also sold hers, taking a heavy capital gains tax.

I had occasion to take only one action in regard to integrity when I was Deputy for Logistics, and that was prompted not by any action of anyone in logistics but by offers and invitations that I knew of others in logistics receiving or that I received myself. All were of the type common in business circles and accepted by the government as proper tax deductions for entertainment. Accordingly, I wrote a letter to each of the Chiefs of Technical Services advising him that I felt it must be assumed that any business expected to receive at least equivalent value for its expenditures on entertainment. Therefore, any courtesy beyond what a military official could personally return or beyond what would be accorded the business on a visit to a military installation might be interpreted as improper influence and was best declined to avoid the risk of embarrassment to the Army.

At a later date, the Department of Defense issued a similar directive on this subject that was widely distributed and aroused considerable clamor. The clamor, however, subsided quickly. I think this was because there was general recognition that it is best for the services to give the appearance as well as the actuality of complete integrity.

Reasons why the problem of integrity recurs.

As long as there are disappointed bidders on contracts and disappointed localities seeking the retention of going operations, there will be challenges against unfavorable decisions. Only so long as the military enjoys an unsullied reputation for integrity can these challenges be effectively restricted to matters of fact or judgment.

Form in which the problem of integrity recur.

There are currently unresolved questions as to what is ethical for a member of Congress and as to what is ethical for a justice of the Supreme Court. Although I feel that the military has a much clearer idea of what is unethical for a logistician, I doubt that it is possible to express it in precise terms. I think there will, therefore, always be a grey area in which interpretations will be required.

CHAPTER 6

Maintenance of Materiel

The supply of repair parts is so critical to effective maintenance that I prefer to deal with both repair and maintenance together.

The maintenance of materiel is a tremendous logistic burden. As already discussed, to reduce this burden some effort has been directed toward improving the reliability and durability of equipment. Still more important is the command supervision that should be given to care in the operation of equipment and to preventive maintenance. I have not dealt with command supervision at length because it does not require analysis, it requires attention. When we had horses in the Field Artillery, the horses were taken care of before the men. If the grade was steep, the cannoneers dismounted and pushed on the wheels; if the footing was bad, the men did some roadwork before the horses were driven across. Inanimate equipment suffers as much from abuse or neglect as did horses but tired men will abuse or neglect equipment unless command supervision is continuously and forcefully exercised.

Making a plan for the maintenance of any complex item or group of items, such as motor vehicles, tracked vehicles, radios, or helicopters, is an involved process requiring consideration of many factors. I have dealt with some of the more critical factors individually and have gathered them into this chapter. They are standardization, repairable versus throwaway assemblies, quality of rebuild, in-theater versus out-of-theater repair, and requirements for and management of repair parts.

Standardization

Commercial practice reflects the teachings of experience and is generally a reliable guide for the military except when conditions encountered in a theater of operations differ greatly from those in the civilian economy. Today, the era of the crossroads mechanic is pretty well over. Stocking repair parts for many different makes of one end item is too great a burden to be economically practicable, as is keeping on hand the many special tools required for different makes and training repairmen for the special operations peculiar to different makes. As a result, automotive service departments specialize in one make of car, stocking repair parts for that make and hiring factory-trained mechanics. Similarly, many manufacturers of electric appliances now have repair agencies for their own products widely distributed. Some firms, particularly those producing computers, only lease their equipment while retaining control of service, repair, and parts.

The military has little time to train repairmen for many makes and models of its equipment, and cannot carry a great variety of special tools and large stockages of repair parts and still be mobile. As a result, standardization is a virtual military necessity.

Along with its manifest advantages, standardization carries certain disadvantages that always require consideration. Standardization of a complex item sometimes means procurement from a very few sources or even a sole source. In an emergency this kind of procurement limits the military's access to the rest of industry until other sources have converted to the production of the standard item. This transition can be difficult. During the Korean War, although a military arsenal could successfully manufacture steel cartridge cases and lent every assistance, there was a long delay before industry could duplicate the process satisfactorily. A similar delay has occurred in the manufacture of rifles. And, as a third example, we once had a number of rear axle failures on a standard truck made by a second manufacturer. The difference was finally found to be in the hardness of the steel used in the rear axle housing. The first manufacturer used steel of a Brinell hardness of 10, but this did not appear on the specifications because his standard factory practice was to use steel of that hardness for any rear axle housings. The second manufacturer used a softer steel.

The use of a sole source results in the loss of competition on subsequent annual buys. It denies to the military the use of proprietary processes and items of other companies that might have been found to be desirable. It reduces incentive toward continuous product improvement. It causes large amounts of funds to go to a major firm as the prime contractor and freezes out small firms except as they may be selected by the prime contractor as subcontractors for parts of components.

This last is of particular concern to congressmen, most of whom receive appeals from constituents seeking government business. Where the constituent is an unsuccessful bidder he seldom has much of a case. But the story is different when, because a prime contractor prefers to negotiate with his usual subcontractors, he has not been allowed to bid at all. Congress generally feels that government procurement ought to be open to bidding by all qualified producers not only for the prime contract but also for subcontracts. Prime contractors, on the other hand, feel that, if they are to be responsible for the end item, they ought to be allowed to negotiate with subcontractors with whom they have dealt before and in whom they have confidence. The solution most acceptable to Congress is to have the military break out the maximum practicable number of components, buy them through open competition, and provide them to the prime contractor as government-furnished equipment. This has the disadvantages of requiring the government to administer many more contracts and of making the government responsible to the prime contractor for the quality and promptness of delivery of the components it provides. An alternative, less acceptable to Congress, is to require the prime contractor to secure wide competition in his procurement of components.

The introduction of federal stock numbers facilitated the supply of parts and components already used in more than one end item. Not very much, however, has been done in seeking to introduce the maximum number of standardized parts during the design stage of an end item. The Society of Automotive Engineers, I believe, pioneered this effort in the automotive field with the publication of its handbook. The Army Corps of Engineers has sought to standardize components such as electric motors used in several end items of engineer equipment. A study made by the Logistics Management Institute recommended that design engineers be required to select parts and components from a recommended listing or present a written explanation of why an exception was required. This appears to be the most effective approach to maximum standardization that I have seen. It would require a design engineer to do a great deal of reference work but, it would save him from redesigning parts and components that had already been designed, and it would produce a degree of standardization as yet unseen in the American military.

Although standardization is generally accepted as a virtual military necessity, it is not always practicable. The greatest difficulties in achieving standardization have been found in engineer-type equipment. Most ordnance is required only by the military and is therefore designed, developed, and produced for the military. Engineer equipment, on the other hand, is largely commercial-type equipment and can be bought off the shelf. Since manufacturing plants must be tooled up to produce specially designed ordnance equipment, production runs large enough to supply the requirements of the whole Army are normal. Engineer-type equipment is produced by many relatively small producers, none of whom has the capacity to produce the full Army requirements without giving up its commercial market and becoming a captive industry. So, until there is a degree of concentration in the engineer equipment industry similar to that in the automotive industry, standardization will probably continue to be subordinated to maintaining a healthy competitive engineer equipment industry.

The problem of the need for standardization of military engineer equipment has continually recurred. One solution, which I supported strongly, was to standardize by theater. Thus Europe might have one make of ¾-ton shovel, Vietnam a second make, Korea and Japan together a third make, and the United States several makes. Since access to repair parts is easier for equipment in the United States, makes of engineer equipment of which the Army has few items are best kept here.

Standardization by theater has the disadvantage that troops moving from the United States to a theater have to be re-equipped with items of a make different from that with which they trained. Also, the quantity and diversity of repair parts that must be stocked in U.S. depots is increased. Accordingly, theater standardization as policy is a substitute to be used only when overall standardization cannot be attained.

Reasons why the problem of standardization recurs.

The problem of standardization recurs because of the basic conflict between the desirability of facilitating supply and maintenance by having only one type of each end item and the desirability of spreading military orders widely through industry.

Form in which the problem of standardization will probably recur.

The problem of standardization will probably recur in the form of a requirement to help solve such difficulties as the reduction of tonnage of repair parts that units must stock, the variety of parts that units must stock, the size of catalogues, the burden that ordering repair parts places on the communications system, the difficulty of procuring replacement repair parts from industry, the training of maintenance specialists, and the variety of special tools that must be carried.

Repairable Assemblies Versus Throwaway

Major assemblies had to be replaced in World War II because of lack of adequate maintenance capability. Often, perhaps usually, no effort was made to repair these major assemblies during active operations. In other words, we were treating major assemblies as throwaway assemblies. As manufacturing processes have become more complex and efficient and as maintenance specialists have become harder to train and relatively more expensive, the trend of the civil economy has turned in the same direction. Throwaway assemblies can be made more cheaply, save the time of expensive repairmen and have greater reliability than repaired assemblies. Throwaway assemblies used in replacement do not even have to be identi-

cal internally to the assemblies they replace; they only need to
match them in form, fit, and function. This greatly facilitates
securing competition in procurement. It also facilitates securing production of small quantities, always difficult to arrange
with the larger companies.

*Reasons why the problem of repairable versus throwaway assemblies
recurs.*

Like standardization, the use of throwaway assemblies is an
important possibility in any effort to simplify supply or facilitate maintenance.

*Form in which the problem of repairable versus throwaway assemblies
will probably recur.*

The possibility of converting repairable assemblies to
throwaway assemblies will arise whenever an attempt is being
made to reduce maintenance effort.

Quality of Rebuild

After V–E Day in World War II our forces in Europe
shipped their best motor vehicles to the Pacific. They left a
large proportion of unserviceable vehicles that the theater had
never had the time and maintenance units available to repair.
They also had a large stock of repair parts. At the same time,
the German automotive industry was idle and could be em-
ployed without cost to the United States. The Ordnance officer
of the European Command successfully put these resources
together in the first great rebuild operation. Unfortunately,
others after him sought to do the same thing without the same
resources and with vehicles that had already been rebuilt once
or even twice. Although they earned praise for economy, there
is considerable question whether their efforts were worthwhile.
Experience since has indicated that only the first rebuild is
economically desirable and then only when circumstances, such
as those mentioned above, are favorable.

Before the Korean War broke out, a rebuild operation for
motor vehicles in Japan was started. The vehicles available
were those left from the World War II campaigns in the Pacific. Already worn, they had deteriorated further because of age

and exposure. Japan had never had much of an automotive industry. There were few repair parts, and the U.S. automotive industry, swamped with civilian orders for new vehicles, was most reluctant to produce parts for vehicles no longer in production. The Japanese, long adept at copying products of other nations without the drawings and specifications, produced repair parts in the same manner. So some repair parts used were new, some were worn, some were unused but had deteriorated in storage, some were rebuilt and some were imitations. The performance of these rebuilt vehicles left much to be desired.

As Deputy Chief of Staff for Logistics, I was never satisfied that rebuilding vehicles was desirable except under the conditions that prevailed in Germany at the end of World War II. Although I was unable at the time to secure conclusive data to support my belief (because vehicles lost their identity in the rebuild process), I did change the policy from rebuild to "inspection and repair as necessary" with the intention of closing out all overseas vehicle rebuild operations at an early date. Since my time, data gathered in studies by the Research Analysis Corporation have indicated that the cost of rebuilding major components such as engines and transmissions is about one-third that of new engines and transmissions but that the rebuilt assemblies last only about one-third as long as the new ones. This would make the cost factor about equal. The reduced reliability, however, so critical in war, and the earlier obsolescence make complete vehicular rebuild, especially overseas, a highly questionable process.

The above discussion relates primarily to the quality of rebuild of motor vehicles, and by rebuild I mean complete disassembly. It does not necessarily apply to other items. Rebuild becomes less questionable when only components are rebuilt, especially when those to be rebuilt are a minor part of the whole end item. Thus, for example, the rebuilding of aircraft engines has long proved economical.

I did not ever have a chance to test the idea of returning unserviceable equipment to the original manufacturer for rebuild. This would be practicable only while the manufacturer was still producing the same item. Assuming the willingness of the original manufacturer, then his more skilled personnel, his factory knowledge of tolerances, and his better availability of repair parts might well make rebuild of some types of end items both economical and satisfactory as to quality.

Reasons why the problem of quality of rebuild recurs.

The factors in influencing rebuild vary for each type of item, so separate study and determination are required for each type of item. Data may well be difficult to obtain for many items, but a rebuild decision ought not be made without some knowledge of the quality to be expected from the process.

Form in which the problem of quality of rebuild will probably recur.

The problem will arise in the form of a requirement for a determination of what quality of rebuild is desirable and practicable whenever a maintenance plan is being made with respect to any item or group of items. Many considerations other than quality will enter, such as availability of maintenance units, availability of transportation, rate of new production, and use to which required equipment is to be put, but of all these considerations the quality of rebuild is probably the one most difficult to assess.

In-Theater Versus Out-of-Theater Repair

In World War II the Army generally repaired equipment as close to its point of use as practicable. This was partially because the most critical shortage in the war was cargo shipping, partially because overland transportation was heavily burdened, and partially because equipment had to be returned to action quickly. Possession of air superiority made it possible to repair equipment well forward without enemy interference. These considerations outweighed the disadvantages of having to support additional men in the theater to perform the maintenance.

In the Korean War the nearness of the Japanese industrial complex, adequate shipping, ready accessibility of the ports to the combat zones, and American air superiority made transportation to Japan for repair easy, got the repair job done in

Japan about as rapidly as it could be done in Korea and saved supporting many maintenance units in Korea.

In Vietnam, on the other hand, few facilities make desirable locations for maintenance shops because every installation has to be protected from guerrillas and movement within the theater is subject to attack. On the other hand, shipping and supplies are plentiful. As a result, some maintenance is done well forward, in division areas, to avoid long movements within the theater, but expensive equipment requiring difficult or extensive repair is shipped to Japan or the United States. The repair cycle for this equipment is long, introducing a requirement for a large pipeline.

For any future sophisticated war there are many new developments that will complicate the determination of whether or not to repair intheater. Long-range missiles will make large repair shops vulnerable targets. Possible enemy air equality or even superiority will make them still more vulnerable. Enemy infiltrators will be able to pass through our lines, more dispersed for fear of enemy nuclear attack, and will make intratheater movement more difficult. Moreover, the vulnerability of shipping to nuclear submarine attack and the vulnerability of ports to missile and air attack will make out-of-theater water transportation more difficult for heavy items for repair. On the other hand, increased availability of air transport should make out-of-theater repair for light items easier even though air and missile attack on our air bases may hamper this movement. These considerations indicate the desirability of spending more money for durability and ease of maintenance of heavy items.

Reasons why the problem of in-theater versus out-of-theater repair recurs.

Many design and production decisions ought to be made in the light of whether the item concerned will probably be repaired largely in theater or out of theater.

Form in which the problem of in-theater versus out-of-theater repair will probably recur.

The problem of in-theater versus out-of-theater maintenance should recur in the form of a requirement to determine the influence that repair policies should have on the design of new items. Thus an item might have to be light enough to facilitate air transport out of the theater, or components most difficult to repair might have to be readily accessible for removal and built to facilitate separate shipment. The problem should recur in connection with the logistic support plan for any theater since fewer maintenance troops will be required if much maintenance is done out of theater, and provision must be made for backhaul and outloading. The problem should also recur in the form of a requirement for the justification for air transport: small, light, and expensive items being repaired out of theater should be shipped by air; complex delicate items should also be shipped by air as the most rapid and least rugged means practicable.

Requirements for Repair Parts

The forecasting future requirements for repair parts for any end item so that they can be stocked in adequate, but not excessive, quantity and identifying and handling the tremendous number and variety of parts required by military units in the field are among the most difficult of logistic problems.

Many approaches have been attempted toward forecasting requirements for repair parts. Estimates based on knowledge and judgment are usually required of a manufacturer before the first production run of an end item. Such estimates are of necessity based on experience with similar items. However, if the end item is completely new, there will, of course, be no similar items by which to judge. Even if similar items have been produced, changes are normally made to correct weaknesses found in them, so old consumption data may not be very good. Some data on repair parts requirements are acquired during engineering and troop tests, but these tests are short and the end items usually operate under relatively favorable conditions such as with good preventive maintenance and prompt repair.

The consumption of repair parts by a new item bears little resemblance to the consumption of repair parts by the same item after a considerable amount of rough usage, aging, exposure, and poor maintenance. I once had the first two helicopters from the production run of a new model run to destruction in an effort to determine the repair parts requirements early, but the experiment was not particularly successful. Continuous use does not result in the same wear as intermittent use, nor does it reflect the effects of deterioration from time and exposure or the effects of a rugged environment and poor maintenance. Tracked vehicles in North Africa in World War II wore out tracks, drive sprockets, and idler wheels at a vastly accelerated rate because sand and oil make a very abrasive compound. Lava ash from an eruption of Vesuvius caused us to experience rapid wear of brakes in Italy, with a resulting critical shortage of brake lining. Thievery of jeeps in North Africa caused heavy consumption of distributor rotors. (Removing the distributor rotor was for a time a more effective protection than removing the ignition key. It ceased to be much protection when those inclined toward larceny or borrowing obtained spare distributor rotors, upsetting estimates of requirements.) Moisture and heat in Vietnam have caused severe deterioration of many parts of radio equipment.

Inability to forecast repair parts requirements accurately accents the importance of gathering and analyzing consumption data promptly after an end item is put into service in the United States and again after it is introduced into a theater. The present reporting system, with some modification, may be adequate, but thorough analysis has usually been lacking. RAC study groups have done outstanding work in extracting, purifying, supplementing, and analyzing existing reports and have been able to produce excellent forecasts of the life of many important assemblies.

Reasons why the problem of forecasting requirements for repair parts recurs.

The size of the initial buy of repair parts, the size of the reserve of each part to be maintained, the size of the final buy when an item goes out of production, and the avoidance of surpluses left over after the end item goes out of the system— all depend upon the accuracy of consumption forecasts made at various periods in the service life of an end item. Since, for the various reasons previously discussed, accurate forecasts of the requirements for repair parts and assemblies cannot be made prior to extensive use of the end item and since adequate forecasts are not being made after use is initiated, the resulting shortages and excesses cause the problem to recur. The successful work of RAC study groups in forecasting the life of important assemblies indicates that such studies should be made of each new model of a critical item in each new environment in which it is operated.

Form in which the problem of forecasting requirements for repair parts will probably recur.

The form in which the problem will probably recur will be as a requirement to modify the reporting system to insure that, for a new and critically important end item, the reporting system will provide adequate data for forecasting consumption of assemblies and parts or as a requirement to analyze the data provided by the reporting system and to forecast the life or consumption rates of important parts, important assemblies, or an end item itself.

Management of Repair Parts

Even if it were possible to forecast requirements for repair parts accurately, there would still be problems in their management. The requirement is for a system that is reliable and rapid yet simple enough to be operated by personnel with little training—this without unduly reducing the mobility of units or even depots in theaters of operations. As in all recurring problems, there are conflicting influences, and modifications can be made by shifting the balance. Thus, complete capability to repair an end item requires that every part of which the end item is made be available for requisitioning. On the other hand, simplicity, mobility, and speed of operations exert influences toward reducing parts available for requisitioning to a relatively small percentage of those that go into the assembly of the end item. Providing interchangeable parts common to several end items leads to storing parts in order by Federal Stock Number. On the other hand, simplicity and reliability exert an influence toward segregation of parts by end item so that in one "master depot" as in World War II or in one section of a depot overseas one can quickly find everything needed for a specific end item. In an Army made up largely of draftees—some of whom must receive their training in a few months in order to perform their service in an active theater within the two years for which they are drafted—and in an Army in which logistic troops must usually be supplemented by local laborers, simplicity is essential. On the other hand, the increasing complexity of equipment developed in an age of rapidly advancing technology makes simplicity difficult to attain.

In World War II, parts were supplied primarily on the estimates of the Chiefs of Technical Services, with theaters requisitioning more when shortages occurred and reporting when excesses were building up. When theaters reached the point of requisitioning for their needs, the usual practice was to requisition parts based on the rates developed by the service chief with little or no adjustment for the variations from those rates caused by conditions peculiar to the theater. In some cases repair parts were furnished in box lots, each intended to provide a balanced supply of repair parts for a specified number of a particular end item for a specified time. With a section of a depot set aside for parts for each important end item, these box lots provided ready restockage. When a new supply point had to be established, a certain degree of balance was assured by sending box lots. When shipments were mixed up en route, they were easy to segregate. When a depot moved, it was easy to send forward a small but fairly balanced stockage. The system was criticized because some types of parts were in excess. Thus, a box lot of ¾-ton truck parts contained, among many fast-moving parts, a steering wheel and a rear axle housing, which were seldom required. This criticism could have been met by changing the contents of the box lots, although it would not have been worthwhile to modify box lots already delivered by the manufacturers.

At the end of the war, we had in Germany a great quantity of repair parts, some of which had been repacked and shipped a number of times. Documentation had been lost or never prepared. It required a major operation, with skilled parts specialists sent from the United States, to sort, clean, identify, and determine the serviceability of these repair parts and return them to stock. If we had not been able to obtain local labor at no expense to the United States, I doubt that the operation would have been worthwhile. There is certainly a case for a policy of disposing of parts not obviously in good condition, not easily identified, or not obviously expensive.

In Korea, a widespread practice developed whereby units removed components from end items prior to turning the end items in for higher level maintenance because they had experienced difficulty in getting these components by requisition. These components were usually held in units not authorized to hold them. Maintenance units had to replace these components, and this caused strange fluctuations in repair parts requirements. Units became overloaded, but this was not serious because of the degree of stabilization that existed in the middle and late periods of the war. The practice was condemned by the technical services but tolerated, if not encouraged, by the combat arms. It was overcome only when maintenance units refused to accept incomplete end items for repair. The best preventive against such practices is, of course, for units to be able to obtain parts and assemblies easily when required.

In World War II and Korea the supply chain paralleled the command chain. Although the impetus for supply was supposed to come from the rear, there was also supposed to be an impetus from the front. Thus, having submitted a requisition, a supply man was supposed to wait a reasonable period. Then, if the item or items required were important, he was supposed to visit the next higher echelon, bringing samples of the required items if identification appeared to be a problem. If he didn't get the supplies there, he went to the next higher echelon and his commander to the next higher commander. It was remarkable how poorly the supply system worked, particularly for repair parts, when this type of follow-up was not used and how well when it was used.

In the early 1950s the Army, in an effort to take advantage of new developments, introduced into its supply system Federal Stock Numbers (to facilitate the use of parts interchangeable between two or more end items), the one-line requisition (to eliminate the delay cause by assembling multi-item requisitions and then breaking them down among depots for supply), electronic transmission of the data on punch cards (to reduce the time from requisition to delivery), and computerized inventories (to maintain inventories up to date without manual inventorying). Under this system, called the Modern Army Supply System, requisitions went to a centralized theater inventory control point, which directed a depot having the item in stock to ship. Thus, a unit in Vicenza tactically based on a line of communications from Leghorn could not requisition on a depot in Leghorn. Rather, its requisition went to Maison Fort in France, which directed shipment from any depot having the item and from which shipping costs were low. This system put a tremendous burden on the theater communications network.

It also broke the clear-cut line of responsibility for supply so that if supply failed to arrive, it was difficult to place responsibility. If radio communications failed or were jammed, the whole system would break down. If a requisition had to go to the United States, it was automatically forwarded to an appropriate U.S. depot for supply without editing, and neither intentional nor unintentional errors were caught. For example, a sergeant in Europe requisitioned an expensive turret lathe. There was no shortage, the lathe was not authorized to his unit, nor did his unit have a special need for one; the sergeant had just always wanted one. It was delivered. Of course, most errors were not intentional. Some were errors in identifying the proper Federal Stock Number or in copying its many digits. These caused erroneous deliveries or delays while the proper number was sought. Other errors were caused by re-requisitioning hard-to-get items. Inventories were maintained at the theater inventory control point and corrected by reports of transactions at depots. Depot commanders had no knowledge of the inventories in their own depots, nor were physical inventories made to verify or correct automatic data processing inventories. Parts in depots were stocked by number. Gone were the days when, as a last resort, a supply man could take in his hand an unserviceable part of a Reo 2½-ton truck for which he had been unable to secure a replacement, go to the Ordnance depot upon which he was based, find in the depot a section devoted to parts for the Reo 2½-ton truck, find the parts man in charge of the section who was familiar with parts for the truck, and get the part.

When I became Deputy for Logistics in 1955, this Modern Army Supply System was in operation. It was working poorly. The problem I faced was whether or not to keep the system and hope that, by modifying the procedures and training the personnel, its advantages might be retained and its weaknesses reduced. The alternative I considered was initially to limit the use of the automatic data processing equipment to the performance of the functions that we had previously found to be essential and so had performed manually—but to perform them faster and with a higher degree of accuracy. Only after this basic system was operating effectively would we expand it to take advantage of any of the capabilities of the equipment to perform the additional functions that it was then performing unsatisfactorily under the Modern Army Supply System. I probably made the wrong decision. I kept the Modern Army Supply System with minor modifications. The system, with some additional modifications, is still in operation. It still works poorly. Even reaction speed, which was supposed to be one of its principal advantages, has never lived up to expectations. The modified system now in operation has been surveyed extensively and many causes of failure and delay have been identified. The lack of simplicity, the impracticability of follow-up, and the probable collapse in the event of communications difficulties appear to me to be the major weaknesses.

If much heavier stockages were held at each level, performance of the supply system could undoubtedly be improved—except in the accumulation of excesses. However, Army units, even depots, must be able to move on short notice, so stocks must be held down to critical and fast-moving items. When I was Deputy for Logistics, we introduced a system that included "prescribed load lists" for the stockage of fast-moving items and relied primarily on cannibalization for the provision of slow-moving "fringe" items. Fringe items, which included most repair parts, were required seldom and usually in lots of only one. In war there are usually many more damaged end items than can be repaired anyway. The parking for cannibalization at each direct support unit of at least one end item of each important type would provide a ready source of slow-moving fringe items. On rapid unit movements, items undergoing cannibalization could be abandoned and replaced at the new location by other damaged items.

In time of peace it was intended that, if no damaged end item were available, undamaged end items would be used. This was supposed to be limited to some extent by judgment. For example, new and expensive end items for cannibalization might be permitted only to general support maintenance units. The reduction in requisitioning was expected to more than pay for the cost of even new items cannibalized in peace. Supply time of fringe items would have been tremendously reduced. This system was prostituted as a result of a misguided interpretation of economy that not only indicated that serviceable or slightly damaged end items should not be cannibalized (leaving no source for fringe items except requisitioning) but even required that items cannibalized should be disassembled and the parts picked up on stock records, this increasing paper work and reducing mobility.

Reasons why the problem of management of repair parts recurs.

Repair parts are in tremendous variety and difficult to identify. Federal Stock Numbers contain many digits and are the source of many errors. The system in use makes follow-up of unfilled requisitions difficult. It places a tremendous load on the available communications. It is not delivering the parts required in a reasonable time, and some not at all.

Form in which the problem of management of repair parts will probably recur.

As combat units continue to be unable to secure promptly the parts they require, the problem will arise in the form of a requirement to redesign the system either in whole or in part to secure greater simplicity, greater reliability, and greater mobility, or any combination of these.

CHAPTER 7

Intertheater Transportation

Overseas operations give rise to many transportation problems. Although some of these problems are primarily the responsibility of the Navy or Air Force, almost all of them have important implications for the Army since it is the Army that must be transported and supported overseas. The Army should therefore be prepared to influence their solution influence on their solution. The areas in which I have particularly noted that problems recur are:

1. The speed of initial deployment overseas.
2. The balance between sealift and airlift.
3. The best method of moving seaborne cargo ashore.

Speed of Initial Deployment Overseas

During and since World War II, and particularly since we became interested in Southeast Asia, a great deal of thought has gone into speed of initial deployment. The concern has been both speed of deployment to NATO in Europe and speed of deployment to Asian areas on the fringe of the Communist bloc. There has been two assumptions: that the early arrival of U.S. forces in the objective area might well be critically important and in some cases decisive and that a smaller force delivered quickly might accomplish as much as a larger force delivered later.

Before an answer can be given to the problem of how rapidly we can place a force of a specified size in a specified area, many questions must be resolved. For example, an important question is whether the movement is primarily for the psychological effect of presence or if the force is expected to engage in heavy combat. If the movement is for a psychological purpose like showing the flag, the force could be moved with little more than individual equipment, but if the force is expected to fight, it should be moved with full combat equipment and reserve supplies. The second type of movement requires, in terms of tonnage per man, many times the amount of the first. Normandy was at one extreme end of the scale; we knew we would fight a first-class enemy. The initial movement to Korea was less extreme; we expected a much less determined and competent enemy and did not take the time to complete the organization and equipment of the troops. Lebanon was further down the scale. Although primarily psychological, the force went with most of its equipment and some supplies, which added to the psychological-political effect, demonstrating a stronger determination than a mere landing would have indicated.

Once the size and composition of the force and the mission it is expected to accomplish are decided, there are many preparatory measures that can be taken to increase the speed with which deployment can be accomplished. It is entirely practicable to take preparatory measures for a very rapid deployment, but most of them must be initiated well in advance, and their costs are high. For example, we are now withdrawing troops from Vietnam. If we wished to redeploy them to Vietnam, they, their equipment, and their necessary supplies would be available and could be moved in record time because the line of communications between the United States and Vietnam is operating at a large capacity. However, it required serveral years to establish this capacity. When the question arises as to how long it will take to deploy a specified force into an overseas area, the problem becomes one of determining what preparatory measures need to be taken, at what time and at what cost, and how fast they can be accomplished. The preparatory measures that need to be taken vary with the geographic location of the objective area, with the situation in the objective area, with the strength and condition of the troops to be sent, with the degree of governmental support behind the operation and with many other influences.

The prepartory measure that probably requires the longest time is the provision of the specialized transportation that would in many cases speed up the deployment. The development and production of the C–5A aircraft, the roll-on, roll-off ship, the fast-deployment logistic ship, and the larger types of landing craft are examples. The economics of peacetime use can limit the provision of these aids to rapid deployment. To the extent that they cannot be economically utilized in peace to move military cargo, they compete for funds with combatant aircraft and ships. The case for the C–5A is particularly strong because it is economically usable in peace and it avoids the threat of the nuclear submarine in war. The case for the roll-on, roll-off ship is good because it moves cargo in peacetime almost as cheaply as commercial shipping (making up for carrying less than normal tonnage by requiring far less than normal time to load and unload). The case for the fast-deployment logistic ship is less strong because it is a very special-purpose ship that is less economical to operate in peace and therefore competes for funds with Navy combatant ships. The same applies to the larger types of landing craft.

Another preparatory measure that may require a long time is the construction or improvement of airports and seaports. The airports and seaports of the United States are adequate for almost any level of overseas movement, except as regards outshipment of ammunition, bombs, and missiles, where safety provisons for the local civilian population cannot meet Coast Guard standards. Adequate facilities are found in few potential objective areas. Improvements are virtually always required. When no facilities exist, construction is a matter of months or even years.

With airports well spaced across the Pacific, along the southern fringe of Asia, along the Mediterranean, and across the Atlantic—all in use by the Military Airlift Command—the United States now has some capacity for air movement virtually anywhere in the world provided there is an adequate air base in the objective area. Two questions, however, normally arise: one is the diplomatic question of authority to use airports and to make overflights of foreign territory for the purpose of the specific movement intended; the other is the question of the improvement, maintenance, stockage, and operation of the airports to be used.

Other nations can be very sensitive to the purposes of flights over their territories and to the use to which even U.S. operated bases in their territories are put. Prior agreements in which the specific purpose of the intended movement was not contemplated are of doubtful value. Refusals are common. For example, in the Lebanon crisis of 1959, Greece, although our ally in NATO, refused permission for U.S. troops going to Lebanon to overfly Greek territory. When we first sent troop detachments to Vietnam, India, although a recipient of U.S. aid, refused permission for U.S. troops to overfly Indian territory. Even if clearance is obtained, any air base that must sustain unusually heavy traffic has to heavily stocked with fuel and repair parts and manned with personnel for servicing and repairing aircraft and maintaining runways. This is a major requirement. Just how major was indicated by a computation made when we first became involved in Vietnam. This computation indicated that it would require as much time to make these preparations at the necessary air bases as it would to transport the first troops to Vietnam by water.

Since most commercial cargo moves by sea, existing seaports usually have far more capacity than existing airports. More way stations are usually available and already fairly well stocked with fuel. Ships can either make the round trip without refueling or can refuel at some port en route without encountering the sensitivity about airports. The high seas are open to transit, so diplomatic clearance is required only if movement is to go through claimed territorial waters or through the Suez Canal. If port capacity in the objective area is inadequate, the need for new construction can usually be temporarily avoided by using landing craft, although the initial movement will usually be slowed down by the need to move the landing craft to the port of debarkation.

Aircraft and ships are scattered all over the world. Few active ships are empty and available in home ports. The assembly of shipping for a large movement is a matter of several weeks or even months. We do have a large number of older cargo ships in storage, but the time required to put such ships back in commission is considerable. Aircraft are more readily available than ships because their turnaround time is so much less. Both ships and aircraft have important commercial missions that cannot be suddenly discontinued without serious disadvantages. In World War II the British Import Program, required to keep the British economy running, was given priority over strictly military requirements. As railway passenger service in the United States declines, our ability to withdraw commerical aircraft for military service will also decline. Further, the aircraft and ships we assemble must be protected. Depending upon the capabilities of the enemy, naval escorts for our ship convoys, air cover for our transport aircraft approaching the theater, and security for our ports and airports in the objective area may need to be provided.

There is always a considerable number of individuals in each unit who are not available for combat service overseas: those sick, those absent without leave, those attending schools, those whose term of service is approaching its end, those facing trial by general court matrial, or those restricted by political measures (such as, at one time, those brought into the Army for service in the Western Hemisphere only). Replacing these men and filling any personnel shortages that already exist with qualified personnel can cause a serious delay. When we sent troops to Iceland before we entered World War II, one of the first requirements was for an Engineer battalion. By transferring men from the other battalion of the same regiment, the first battalion was made ready in a few days. This however rendered the second battalion entirely unfit for overseas service and, when it was called for, a delay of several weeks was required, and even then the battalion was handicapped by an excessive number of replacements. An expeditionary force can be kept ready at all time, but at the expense of personnel turbulence throughout the Army that the Army itself has generally been unwilling to accept.

It is the nature of our democracy that we do not often take, in advance, all or even many measures to permit rapid intervention in any foreign area. This is probably because any overt preparation that has not been publicly debated and cleared is subject to congressional challenge and diplomatic embarrassment. We can therefore keep troops in Europe under NATO, as a sort of extension of World War I; troops in Japan and Korea, as a sort of extension of World II and the Korean War;

and troops in Thailand incident to the war in Vietnam. But in none of the ways mentioned, nor in World War I or Lebanon, did we decide in advance to be prepared to intervene rapidly and take many of the concrete measures that would facilitate such intervention.

We have taken some measures. In an effort to make the best use of airlift in expediting initial deployment, forward stockage of equipment has been used. The maximum use of forward stockage has been made in Europe, where division sets of equipment have been stocked with a view to flying the troops to their equipment. This requires a heavy investment in depot space and in personnel to maintain the equipment in storage and make it ready for issue. It also involves the procurement of two sets of equipment for the same division. As equipment depreciates in storage or becomes obsolete, it must be replaced in both division sets.

A lesser degree of forward stockage has been used when troops are not in the objective area. This has included storing heavy and bulky equipment and supplies such as tanks, trucks, heavy engineer equipment, artillery, and ammunition in an area in which U.S. troops are already stationed that is part way to one or preferably several objective areas. This measure has the great advantage of reducing depreciation and obsolescence through issue to the local troops of equipment from the forward stockage and replacement from the United States.

We have provided, under foreign military aid, military equipment and supplies for the troops of countries in whose interest we might intervene. If such countries are threatened, whether from without or within, they must fight in their own defense; else we are unlikely to intervene. If their troops are not overwhelmed, they can hold seaports and airports open to facilitate our troop movements. If they are overwhelmed, the problem ceases to be rapid intervention and becomes massive intervention. But in regard to forward stockage, another possible approach would be to issue to a threatened country, through foreign military aid, more than enough stocks for its own forces. An American expeditionary force could then borrow back some of the bulky or heavy items. This approach has been considered but, so far as I know, never used.

The speed of movement to an objective area is thus not a simple question of how fast an airplane can fly or a ship can sail. It is a complicated problem in which there are many factors, and these factors can be greatly changed by preparations that require time and money and that reveal our intentions.

Reasons why the problem of speed of deployment recurs.

Each time the deployment of troops to some area of the world is studied, one of the problems raised is how quickly deployment can be made.

Form in which the problem of speed of deployment will probably recur.

The form in which the problem usually recurs is with respect to the degree to which some measure relating to transport means—forward stockage, reserve fleet, air-transportable equipment, for example—will increase the speed of deployment to some critical objective area. The problem may recur just as the overall question arises of how fast military support of a specified size can be provided to some specified area. Perhaps the most general form in which the problem will probably recur will be as a requirement for a methodology, preferably computer assisted, by which the influences on the speed of deployment to any area can be varied among possible values and the effect of any combination determined. Such a methodology, applied in a war game, should give valuable information.

Balance Between Sealift and Airlift

Movement by air overseas is much faster than movement by sea, but capacity is more limited. In commercial practice, air carries overseas something less than 5 percent of freight, water over 95 percent. Air transport can be increased to perhaps 15 percent, but this increase is not a matter of aircraft alone but of building, manning, stocking and operating bases on the route to the objective area and in it. Even when an air line of communications is in full operation, its capacity can be greatly increased only by such measures as moving in additional aircraft maintenance personnel, and in runway maintenance equipment and personnel, increasing stockage of fuel and parts, and perhaps building additional runways, hardstands, and fuel storage.

Movement of resupply by air instead of water cuts order and shipping time less than the difference in transit times would indicate. If an overall order and shipping time by water is forecast at 120 days, a reduction of transit time from 20–30 days by water to 1–2 days by air does not reduce order and shipping time proportionally. Expediting action all along the line is also required if a major speedup in delivery is to be accomplished. This involves such measures as special requisitions, granting of priority, elimination of editing, direct communication to the supplying depot, priority for picking and packing at the depot, lighter packing, air or express shipment to the aerial port of embarkation, prompt loading, prompt departure of loaded aircraft, prompt unloading of aircraft at overseas air bases, and prompt forwarding of freight to destination. This movement can of course be further speeded if the same plane picks up freight at the depot and delivers it to the air base of ultimate destination. But this measure is practical only in special circumstances, not for general-purpose freight movement. The larger and more varied the tonnage to be moved, the less expediting action is workable. The effect of a combination of air movement and expediting action has reduced order and shipping time for Vietnam from an average of 98 days for routine handling and shipment by water to 43 days for expedited handling and shipment by air. Air transport is therefore still basically a supplementary means of overseas transport particularly adaptable to the movement of personnel and high-priority, lightweight cargo.

In every overseas operation, our buildup in the area initially occupied must be faster than what the enemy can accomplish lest we risk early defeat. The time for initial deployment should therefore be reduced as much as possible. Airlift is one of the means of expediting deployment, and its maximum use must therefore be considered in the early phases of any overseas operation. So far as I know, the German conquest of Crete has been the only overseas operation in which the whole force was delivered into the objective area by airlift. This was a Pyrrhic victory. The losses were so heavy that the Germans were never able to mount another such effort. All deployments since then have been made primarily by sealift, with airlift in a strictly secondary role. This does not necessarily hold for the future as the airlift available steadily increases and as more and more airbases are built all around the world. Changed conditions must be evaluated for each operation planned in the future.

In the later phases of any overseas operation, rapid resupply of critically needed personnel and materiel gives airlift a particular advantage. This was the type of use of airlift that the United States has made in World War II, Korea, and Vietnam. The practice has been to have the staff of the overseas theater decide on the utilization of the airlift tonnage from the United States allotted to the theater.

Finally, airlift avoids the nuclear-powered submarine threat, which may cause heavy losses of shipping in any war in which the Soviet Union takes part.

Reasons why the problem of the balance between airlift and sealift recurs.

In the planning for every overseas operation, the problem of how best to use available airlift arises. In preparing the Army's recommendations for procuring ships and aircraft for military lift, the most desirable balance for the more important contingency plans should be considered.

Form in which the problem of the balance between airlift and sealift will probably recur.

The problem of this balance arises in the form of a requirement for recommendations concerning the procurement of such items as the C–5A aircraft and the fast-deployed logistic ships. It arises with every requirement for the preparation of a contingency plan for an overseas operation.

Moving Seaborne Cargo Ashore

Delivering ships to an overseas theater is primarily a Navy problem, but moving troops and cargo ashore is a problem in which the Army has an important and sometimes a primary interest.

Shipping immobilized offshore for unloading is highly vulnerable to submarine attack. Shipping in port offers a concentrated and highly vulnerable target for air or missile attack. Moving troops and cargo ashore in minimum time and with minimum loss is a basic requirement of overseas operations. The speed with which troops and cargo are moved ashore reduces the period of maximum vulnerability. It also reduces the shipping and port facilities required. Matching up ports, ship, port equipment, and port personnel for effectiveness and speed is a critical problem.

Although it is possible in fair weather to make an initial landing on an open beach, ports are an early necessity. If they are blocked, they must be opened; if they are damaged, they must be repaired; if none are available, they must be built. In World War II, parts such as Naples were frequently damaged by our own bombing prior to their capture. If damage by our bombing was light, demolitions were carried out by the enemy prior to his evacuation, as was done at Leghorn. Both examples resulted in extensive destruction of port facilities and equipment necessary to the operation of the port. Piers, railway culverts, bridges, and tunnels blown, tracks cut, cranes destroyed or evacuated, and power generation facilities destroyed. In addition, ports were blocked with sunken ships, cranes, and harbor craft. This must all be considered in advance so that the attacking forces bring with them equipment to replace what has been carried away or damaged beyond repair.

Port construction where no port existed was accomplished in World War II for the Normandy landings by the sinking of Phoenixes and hulks to form harbors on open beaches. Ports were unblocked and repaired in World War II by Navy marine salvage teams and Army engineers working together. These units acquired many of their skills by experience in that war. The Germans became more skillful at blocking and damaging ports and the Americans more skillful at unblocking and repairing them as the Mediterranean campaigns required, in succession, the opening of the ports of Bizerte, Naples, Leghorn, and Marseilles. Opportunities for good training in this area in peacetime are rare.

In World War II, ports were vulnerable targets. At Bari we lost 16 ships, 38,000 tons of cargo, and 1,000 casualties in one air raid. With the development of missiles, ports have become more vulnerable. After World War II, in an effort to ensure dispersion of shipping in port in any future European war, the NODEX exercises were held periodically. These exercises involved selecting locations, dispersed for protection against nuclear attack, along the Garonne estuary and the western coast of France where up to five ships could be anchored or berthed and unloaded. A new location was supposed to be tried out in each exercise to test the adequacy of anchorages, the effect of tides and currents, and the accessibility to roads and railroads, and to determine what should be done in the event of war to make the location suitable for use as a small port.

Ports offered no great problem in the Korean War. Pusan was a fine port and was always in our hands. Inchon and Wonsan were captured undamaged, although Wonsan was heavily mined. In Vietnam inadequate port capacity forced the construction of a new port at Cam Ranh Bay, where the DeLong pier was used to expedite operations. Developed after World War II, the DeLong pier is a prefabricated pier that can be floated into place and then jacked up on its own piles.

Most of the cargo for overseas theaters must be carried in commercial shipping. Special-purpose shipping is necessary in many cases, particularly LSTs and smaller landing craft, but such craft are generally uneconomical to use in peacetime and in reality compete for funds with combatant ships. The LST, however, proved itself in World War II and Korea. It was invaluable at Anzio, when most cargo was delivered in loaded trucks. These were driven ashore, unloaded, and driven back on board the LST, which sailed the same night it arrived. The speed with which an LST can be loaded and unloaded of vehicles make it economical for use in peacetime when short hauls are required as between Japan and Korea.

After World War II the military pioneered the use of an overseas shipping container that, among its other advantages, speeds up cargo handling. The much larger commercial containers now coming into use speed up cargo handling still more. The military can foster this militarily desirable commercial development by using containers for shipping in peacetime.

Vehicles are an important part of military cargo. Their movement requires a disproportionate amount of space for their weight. During World War II, this disproportion demanded the partial disassembly of manufactured vehicles and crating for shipment in twin or single-unit packs. Although this practice made cargo handling less difficult, it required complicated reassembly. The roll-on, roll-off ship was developed after World War II to handle vehicles. It is valuable militarily in that it reduces the time of great vulnerability of shipping in overseas ports in war and avoids disassembly, crating and reassembly of wheeled vehicles.

Speed of unloading can be greatly facilitated by port equipment appropriate to the port to be used. Mobile cranes, floating cranes, rough terrain forklift trucks, harbor craft, and amphibious vehicles are all valuable under certain conditions. In over-the-beach operations, amphibious vehicles such as the amphibious 2½-ton truck (DUKW) in World War II were particularly valuable because they could deliver cargo directly from ships to dumps, avoiding the vulnerable and slow cargo transfer operation that would otherwise have been conducted at the water line. Heavy-lift helicopters promise to be valuable in this type of operation provided the ships are suitable for the operation of helicopters.

Reasons why the problem of moving seaborne cargo ashore recurs.

The success of any major overseas operation depends heavily upon the efficiency and speed with which cargo is moved ashore. Every new possibility that might improve our capability in this sensitive area should be carefully evaluated. The problem will recur whenever an overseas operation is planned and the adequacy of the means available to get cargo ashore rapidly is considered. It will recur each year when the budget is developed and the decision is made concerning what landing equipment should be procured by the Army or recommended by the Army for procurement by other agencies. It will recur whenever consideration is given to the development or procurement of new types of ships or new types of port equipment.

Form in which the problem of moving seaborne cargo ashore will probably recur.

The problem will recur in the form of a requirement to determine the equipment that should be provided for a specific overseas operation or that should be procured and kept on hand as a general-purpose reserve. It will recur as a requirement for recommendations as to any concepts of ship-to-shore cargo movement or new types of cargo ships or new pieces of landing equipment.

CHAPTER 8

Production for War

For many years the Army has supported the proposition that supply readiness for war should be provided partially by reserve stocks on hand, partially by production lines in operation, and partially by production lines in mothballs. It also emphasized that the size of the reserve stocks ideally must be great enough to fill wartime requirements until the rate of production can reach the established rate of consumption. The Army Staff has also favored the assumption that any war will start without a preliminary period of strained relations in which production can. This provision of reserve to last from D-day until production equals consumption, called the "D–P Day Concept," has usually been considered theoretically reasonable by most national administrations. But the initial cost would be so great, the cost of storage and maintenance would be so high, and the effect of obsolescence would be so severe that no budget has ever provided for more than a very small fraction of either the reserve stocks or the production capacity required. It has therefore been necessary for the Army to concentrate its efforts on securing the greatest practicable degree of material readiness from the funds made available.

Some of the more important questions involved in arriving at an overall plan of how to achieve the best possible degree of materiel readiness include:

1. What items should be provided for by stockage or production capability?

2. What production lines should be maintained in active operation and what reserves should be stored and maintained?

3. What is the vulnerability of the production base to nuclear attack, and what are the plans for recovery from the results of a nuclear attack?

Items To Be Provided for by Stockage or Production Capability

The Army has usually been able to have its requirements met for the initial equipment and peacetime training of its active units. It has never had its supply requirements met for a possible future war, so it has to determine those items for which to make a substantial provision and those items for which to make little or no provision. This decision has to be made with respect to each specific item. When I was in logistics in the Department of the Army, I had a study made, and kept alive by frequent revision, that evaluated various factors in regard to each important item of supply in order to reach a determination as to the degree of readiness that should be provided for that item and how it should be provided. The more important factors were criticality to combat, rate of obsolescence, rate of depreciation in storage, availability of commercial substitutes and cost. If an item was highly critical to combat, a substantial provision was made for it. The other factors influenced how the provision was made.

Weapons and communication equipment were high on the list. Administrative and special-purpose equipment were low. For example, tanks were high because they were among our principal items of fighting equipment. Armored personnel carriers were a little lower because, although they brought the infantry to the battlefield fresh and gave some protection against enemy fire and radioactive fallout, infantry could move without them. Observation aircraft were high; transport aircraft, relatively low.

Most newly developed items in areas where the state of the art is changing are subject to extensive modification as use in the field brings out weaknesses and possible improvements. Such items, if bought in great quantity, quickly become obsolete and therefore, to my study concluded, were not to be bought to be placed in reserve. Rather, reliance to replace consumption was to be placed on expanding in war production lines that had been kept operating slowly in peace. Helicopters and missiles were in this category.

For items critical to combat that depreciate rapidly in storage, reserve stocks were to be limited to quantities that would be used up in peacetime training and reliance was to be placed on production capacity. Dry-cell batteries were an example.

Although not as suitable for field use and not as standardized as military items, many commercial items can be used satisfactorily as substitutes, particularly in training. Similarity between commercial items and the corresponding military items makes the conversion of manufacturing facilities to wartime production easy. Large reserves of such items were not to be procured. General-purpose cargo vehicles were in this category.

The cost of individual items is considered before they are standardized and included in Tables of Organization and Equipment. The cost of an individual item entered into the problem of providing equipment readiness primarily as an influence on how high a level of readiness was to be provided and by what means it was to be provided, i.e., by stocks in storage, by production facilities on standby, or by a production line in operation. Thus medium tanks, which were among the items most critical to combat, had to be provided for. They could not be bought for stockage in great quantity because of cost. Because of the threat of obsolescence, it was desirable to depend as much as practicable on a going production line. We had equipment for three production lines. We sought to keep one line as thoroughly modernized as possible and operating at a very low rate. We sought to keep the others as modernized as possible but in storage. Modernization was accomplished for all three lines through an annual contract. When the operator of the going production line lost the competition for the next annual contract, his production line was placed in storage. The new contractor was required to modernize and activate one of the production lines that was previously in storage.

Reasons why the problem of items to be provided for by stockage or production recurs.

As new items are developed or new production methods are devised or as changes arise in the probability of occurrence of various possible future wars, the criticality of specific items to combat changes, as does the justification for their stockage and production. All should therefore be redetermined.

Form in which the problem of items to be provided for by stockage or production will probably recur.

The problem recurs during development of each annual budget in the form of a requirement for a decision that specifies the items for which funds will be sought, whether for the provision of reserves, for the maintenance of reserves, or for ensuring industrial readiness.

Production Lines To Be Maintained in Active Operation

The requirement for going production lines and for materiel reserves varies greatly among the three military services. It takes so long to build new combatant ships that the Navy must be maintained basically as a force in being, since it has to fight any war, except the longest ones, primarily with the ships on hand at the beginning of the war. Army equipment, on the other hand, takes a relatively short time to produce, so the Army can be, and is, basically a mobilization base. The Regular Army is not usually large enough in peacetime to fight even a brush-fire war. It has expected to be expanded after an emergency arises. While it is being expanded, materiel can be produced. The Army therefore depends for materiel primarily on production after an emergency arises. Heavy aircraft take more time to produce than Army types of materiel but less time than ships. The Air Force is in between the Navy and Army. It is primarily a force in being but must rely more heavily than the Navy on new production after an emergency arises.

Since the Army is going to be expanded after an emergency arises and equipped primarily from production initiated or expanded at that time, it is important that there be at least one going production line for each critical item. This production line should be frequently modified so that it is producing the best model that has been developed of that critical item. The production line itself should be thoroughly modernized so that if an emergency arises that requires the establishment of additional production lines, industry has an example to copy and experienced personnel to advise and assist in installation and initial operation.

In the late 1950s great political pressure was exerted to eliminate the Army's arsenals and rely entirely on industry. The Army's response was based primarily on the the fact that there are many items critical to combat for which there are no close commercial counterparts and for which production is not kept going continuously. For such items there is no "captive" industry. When new production is needed after a period in which there has been none, many difficult problems arise. The previous manufacturer may be producing some other important item and thus unable to undertake the new production, or a competitor may win the new contract. In such cases the new manufacturer has to depend upon records and the cooperation of the previous manufacturer. Even with the best of cooperation, which is uncertain, there will be some loss of know-how because of inexperience, new personnel and confusion in the transfer of records. There is often the question of the proprietary rights of the previous manufacturer. Regardless of whether the previous manufacturer or a new one wins the contract, there is often a major time loss because the manufacturer starts to redesign any equipment or procedures that require change because of modification in the end item or because of the introduction of new manufacturing processes only after the awarding of the contract.

For each important item not in continuous production by a commercial concern, the Army needs its own flexible facilities that can do research and development work and also maintain in their personnel the knowledge of how to establish an efficient production line or, better, facilities that keep in operation a production line of limited capacity producing the best item, with the best production equipment, with highly competent experienced personnel. New manufacturers then have access to an operating model and to experienced personnel who can assist them in initiating production of the required item. The

Army's solution to the problem of prompt procurement in an emergency relies on its arsenals in conjunction with such commercial production lines as it can keep in operation.

During the period while I was Deputy for Logistics, it was then Army policy to spread production for the initial provision of any new item over five years or more, so as to keep production lines running, although slowly, and to introduce improvements into going production lines as soon as the responsible Chief of Technical Service felt they were desirable and worth the cost. The object of this approach was to have as many critical-item production lines as possible going before any emergency and to have them producing the best possible item. When it became necessary to expand production to the maximum, we would thus already be producing the item we wanted and there would be no delay to work out manufacturing procedures or to test the end items, as improved, before expansion of production could be accomplished. Admittedly, contracts negotiated with the more competent manufacturers were preferable to wide-open competition. Also, it was accepted that some contractors would endeavor to increase their profits by negotiating high charges for engineering changes. Our protection against excessive charges was the judgment of the contracting officer and the necessity for contractors to protect their reputations in order to secure future contracts.

The conflict that is continually waged is that of economy against quality. Voltaire's comment that "the budget is the enemy of the good" may apply in many circumstances, but it is not valid for military materiel, as I sought to bring out in my discussion of cost-effectiveness in Chapter 5, above.

Reasons why the problem of production lines to be maintained in active operation recurs.

There is little argument as to the desirability of having production lines in operation at the outset of an emergency. The question is, How much is it worth to have a production line in operation and modernized? This question arises whenever the production of an item is planned and whenever an engineering change is considered.

Form in which the problem of production lines to be maintained in active operation will probably recur.

The problem will probably recur as a requirement for a recommendation as to how the procurement of any one of the items particularly critical to combat should be planned. It is sure to recur when there is some threat to the security of the United States strong enough to change the accent from cost reduction to increased military capability.

Storage and Maintenance of Reserves

In addition to the initial cost of reserves, the costs of storage and maintenance exert a strong influence against the buildup of heavy stockages. Few items can be stored without deteriorating, and many require a considerable amount of care but still deteriorate. Canned goods deteriorate. Fabrics deteriorate. In spite of the most careful packing and storage, both the propellant and the high explosive in conventional ammunition deteriorate. Mechanical items of all types deteriorate, especially when two parts made of dissimilar metals touch and electrolysis takes place. Rubber-composition parts and rubber insulation on wires deteriorate. The rate of deterioration of some mechanical items is slowed if they are operated occasionally. Rubber deteriorates more slowly if it is stored in nitrogen. Many items will deteriorate more slowly if they are stored under conditions of controlled temperature and humidity. Generally speaking, various measures can be taken, each at a certain cost, to slow deterioration, but some deterioration still takes place. This indicates the desirability of rotating stocks, that is, issuing the oldest first.

During the period of storage, obsolescence also takes place. The rate of obsolescence is particularly critical where the state of the art is changing rapidly, such as for missile and electronic items. The cost of keeping reserves on hand, then, is the cost of storage and maintenance in storage plus the reduction in value that takes place because of deterioration and obsolescence during the period of storage.

Whenever a long-range plan for the management of an improved model of an item is made, a decision is required as to how the new model and the older models will be distributed. One solution, used in the past, has been to put the new model in the hands of troops and keep the older models in storage. The new model will enhance the combat capability of the troops, but this use will also cause the new items to be worn out in training. The older models kept in storage will then have to be used in war. Another solution has been to issue the older models to troops and put the newer models in storage. Under this solution, the best equipment suffers from deterioration in storage, and the troops never get their hands on it in peacetime except for familiarization. Weaknesses in the new model are not discovered as quickly as when it undergoes heavy use. A third solution has been to issue two sets of equipment, one for training and one for familiarization and use in war. This places a heavy maintenance burden on the troops.

Those who plan for the management of an item soon to go out of the system must decide about continuing its maintenance. One solution has been to repair unserviceable items promptly and put then back into service until they cease to be economically repairable. The units so equipped have usually been those of the National Guard and Reserve that would not be called early in an emergency. This practice permits some of the newer equipment to be stored in reserve. Another solution has been to place unserviceable items of older models in storage unrepaired, as part of the war reserve, with a view toward repair and issue in the event of war. These items are then disposed of unrepaired when they become obsolete. This saves limited maintenance funds and reduces the investment in reserves but keeps obsolescent materiel longer and increases the risk of having to use it in war.

Reasons why the problem of storage and maintenance of reserves recurs.

The above discussions are far from complete, but they should convey the realization that there are many conflicting influences that bear on the acquisition and storage of materiel for war. These influences vary for different items and under different military situations. Whenever any of these influences change, the problem of what to do about storage and maintenance of reserves recurs.

Form in which the problem of storage and maintenance of reserves will probably recur.

The problem will probably recur as a requirement either for a general policy for the acquisition, storage, and maintenance of war reserves or for a recommendation with respect to a specific critical item.

Industrial Security and Recovery From Nuclear Attack

The most critical external threat to the security of the United States is an intercontinental ballistic missile attack by the USSR. The possible effects of a variety of such attacks and the possible rate of industrial recovery from them has been studied to some extent. Civil defense measures have been studied, and some few small steps have been taken to reduce losses and damage from nuclear attack. Our reliance, however, has been placed on our capability to survive a nuclear attack and launch a decisive counterattack the "second strike." To this end, protection has been provided for our nuclear missile systems, but not for our industrial capacity to produce them.

As the USSR increases its nuclear capability and its anti-ballistic missile capability, the credibility of our second-strike capability is reduced. The problem of what the United States should now do if it is decided that a nuclear preponderance in our favor will not be maintained therefore arises.

As our own ballistic missile defense is further developed and as other defensive measures are discovered, the overwhelming effectiveness of a nuclear attack will be lessened and the probable duration even of an all-out nuclear war increased. This, in turn, will increase the importance of production after such a war starts and will focus more attention on industrial security and recovery from nuclear attack.

Reasons why the problem of industrial security and recovery from nuclear attack recurs.

As the effectiveness of the various nuclear weapons and countermeasures fluctuates, the requirement for restudying measures for industrial security and recovery will recur with each important fluctuation.

Form in which the problem of industrial security and recovery from nuclear attack will probably recur.

The problem will probably recur as a requirement for new studies, in the light of changed circumstances, of the probable effects of nuclear attack on the United States, of desirable civil defense measures and of desirable measures to facilitate industrial recovery. The emphasis will likely be on a more prolonged struggle than previously assumed, possibly with the accent on fostering a capability of continuing the war with materiel produced after the initial nuclear exchange.

CHAPTER 9

Lessons Learned in Logistics

In previous chapters I have reviewed various logistic problems that I have observed recur over the years. The specifics of these logistic problems, dealt with individually, do not seem to me to give an adequate conception of the general conclusions that I have reached as a result of my experience in logistics. I have, therefore, added this chapter as a sort of summary, presenting in it the lessons I have learned largely from dealing with the recurring problems previously discussed. Where the basis for a lesson learned has not been adequately brought out in the discussion of recurring problems, I have sought to add it in this chapter.

There is not much that is new to any trained logistician in the statements of lessons learned that I have included. There seems to me, however, to be a good deal in them that has been forgotten or disregarded in the years since World War II when the accent has been on economy and efficiency in peacetime operations as distinguished from preparations for effective operations in war.

There are a few hard and fast rules for the logistician. Probably no one realizes this better than I from the effort to develop some general conclusions from my own experience for this chapter. I trust that those I have developed will be viewed by readers as valid under most of the varied conditions I have encountered in the past but as needing to be confirmed or modified if widely differing conditions are met in the future.

Lessons Learned

1. Among the important requirements for the effective re-supply from the United States of a theater of operations in war are:

a. A flow of material from new production that, as soon as practicable, will become equal to the expected rate of consumption by all theaters of training in the United States and by our aid to allies.

b. Reserves in the United States large enough to fill fore-casted theater requirements until the rate of production can catch up with the rate of consumption and to provide a "surge tank" from which to fill unforecasted theater requirements until production can be adjusted to meet them.

c. A continuous flow of materiel expenditure reports from the theater which show cause of loss or reason for expenditure so as to permit the updating of replacement factors and a recomputation of the forecasted rate of consumption on which both new production and reserve stockages should be based.

It is to be expected that actual average consumption in a new theater will vary widely from that computed with replacement factors developed before war starts. It is to be expected that actual average consumption in an established theater will vary considerably even from that computed with replacement factors developed after theater experience has been accumulated over several months. It is to be expected that actual monthly consumption will vary widely from actual average consumption. Experience in World War II indicated the need for a two-month reserve in the United States even after the rate of production had reached the rate of consumption.

2. The initial equipment and supply requirements of a theater of operations in war are best determined in advance through planning, or better, war-gaming a contingency plan through at least six months of active operations. This provides a basis for necessary estimates of:

a. The time that will elapse before the resupply pipe line from the United States is in full operation.

b. The requirements for equipment not carried in the Tables of Organization and Equipment.

c. The types and expected intensity of operations by month.

d. The level of the theater reserve required to assure balance in distribution and to tide the theater over any probable interruption in the movement of resupply from the United States.

e. The requirements of allies who are to be supplied from the United States.

f. The resources available in the theater that should be exploited.

3. The requirements of a theater of operations in war for ammunition vary between a rather indeterminate minimum needed to strike critical targets effectively and a maximum limited only by the durability of weapons and the physical capability of personnel to operate them. In any major war our raw materiel resources, our production capacity, and our transportation will dictate a limitation on the quantity of ammunition produced. The criticality of opertions being conducted will dictate the allocation of ammunition to theaters and the rationing of ammunition within theaters. In any minor war the desirability of reducing our own casualties may cause ammunition allocations to a theater to approach the maximum.

4. Local wartime procurement in an overseas theater of operations saves time and transportation, conserves the resources of the United States, helps keep the local economy operating, and permits a supported nation to make a greater contribution to the common cause or requires a conquered territory to pay a part of the cost of the war. Local peacetime procurement in overseas areas stimulates the local economy, helps establish quality control in the industry of underdeveloped nations, and teaches our own procurement personnel how to contract with foreign industry.

5. Well-trained logistic troops are a critical requirement in war at the opening of a theater of operation. Without them, even if there is no fighting, unloading of shipping will be delayed and supplies will accumulate unsorted and unidentified

and therefore unusable. In addition, if there is fighting, communications will be poor; the distribution of food, ammunition, and POL will be slow; casualties will be inadequately cared for; and maintenance of materiel will be ineffective. Peacetime reductions in logistic troops should not be allowed to eat into the trained logistic troop requirements for the first three months of any theater of operations whose opening might be required on very short notice. Beyond this, logistic troop requirements may be compromised if necessary to provide trained combat troops.

6. Since smaller logistic troop requirements allow more combat troops in the field, continuous efforts must be made to reduce logistic troop requirements for a theater of operations. Important among the many methods that should be considered are: simplification of distribution by broader use of containers with standard content; improved reliability and durability of equipment; reduction of fuel consumption; use of local labor; use of transportation to support shorter evacuation policies, to return unserviceable equipment to the United States for repair, and to reduce ground lines of communications; reduction of the maintenance load by making more components "throwaway" instead of "repairable"; and organization and training of allied logistic troops.

7. An operational concept should be immediately followed by a transportation capability study. If the transportation system will support, or can be developed in time to support, the forces necessary to carry out a contingency plan, the rest of the logistics scan usually be brought into line.

8. The logistic purpose served by planning is to permit the initiation of action to meet logistic requirements early enough so that they can be provided without upsetting the orderly organization and training of logistic troops or the orderly operation of production and supply of materiel. It is not to be expected that any plan prepared far in advance will be executed without modification or even major revision. It is to be expected, however, that the logistic resources provided to support the original plan will meet the mass of the requirements

for the final plan. The provision of any unforecasted requirements can then be expedited. The more accurate the original plan, the fewer will be the unforecasted requirements. The fewer the unforecasted requirements, the more they can be expedited without inflicting confusion on other areas.

9. The best way to prepare a good contingency plan is to war game it over and over. The combination of two contestants and an umpire will help to provide against weaknesses being overlooked or strengths being overestimated. In successive games the various lines of action, both friendly and enemy, can be tested against each other and the strongest developed.

10. The price that must be paid for good personnnel management is the detailed and continuing personal attention of individuals in high positions. These individuals seek to obtain and retain competent personnel to assist them in discharging those responsibilities; their prestige and authority is adequate for the tasks of recruiting placement and career guidance; and their experience and integrity make their decisions sound. What is needed is some adequate method of performing the personnel functions once performed by the Chiefs of Combat Arms and Technical Services as well as a method of bringing the interest of each member of the General Staff to bear on the careers of senior personnel in his field.

11. The proper reward for competence is increased responsibility. The satisfaction that a good man derives from his work comes fom the feeling that his abilities are being fully utilized on important work. The logistics field is more flexible than the command field, where promotions, pay, and decorations usually follow the assignment of responsibilities, and where reclassification and demotion usually follow inability to perform adequately in combat. There are always routine logistics assignments to which mediocre personnel can be relegated, but there are also challenging assignments crying for outstanding people to fill them.

12. In an age of great technological progress and resulting specialization, the military needs channels by which informa-

tion and guidance relating to a specialty are passed up and down through individuals at each echelon who understand the specialty. The old "technical channels" served this purpose. Some similar channels for the logistics field are needed.

13. Experience has proven that the Department of the Army, with its worldwide responsibilities and broad policy functions, is not the right level to edit requisitions, distribute them to proper agencies for supply, follow up on availability, call forward shipments, arrange for expedited action in shortages, implement priorities for shipment of cargo set by the theater, ensure that cargo documentation is forwarded in advance of shipments, and take measures to replace lost cargo. A monitoring agency is required. Without such an agency the overseas theaters have no single point of contact in the United States to deal with supply failures. Just as a project officer supervises and expedites the production of a specific item, so such an agency (once the Overseas Supply Division at a port of embarkation) should supervise and expedite the supply of a specific theater.

14. The degee that resources are wasted in war can be limited by measures taken before excesses and surpluses accumulate. The following measures are basic:

a. Both the fact and the cause of a materiel expenditure must be reported by theaters of operations. This furnishes the basis for an intelligent estimate of future requirements. The Army's recently developed COLED–V reporting system, to be discontinued in 1970, would meet this requirement for ammunition if reinstituted before an overseas theater is established.

b. Theaters of operations must report overages as well as shortages. Experience has indicated that, particularly in the early phases of an overseas operation, this is neglected. The whole Army needs to recognize that the buildup of excesses in a theater of operations results in a corresponding excess of reserves and production in the United States and that the resources so used should be converted to produce something needed.

c. Requisitions from a theater of operations must be edited by an agency independent of the theater and removed from the pressure and confusion of active operations. Errors should be caught before supply action is taken, and excessive requirements should be challenged and, if not justified, then reduced.

15. Measures to redistribute excesses and use surpluses are of questionable value. The important objective is to prevent their accumulation. The time, effort, and cost of storage, correspondence, transfer, and bookkeeping are such that it is usually economical to keep material that is locally excess until it is used and to dispose promptly of material that is surplus after weighing the cost of holding it against any increased return that may be obtained by delay to seek a better price.

16. Decisions should be made at the lowest level at which all the important factors bearing on a problem can be adequately weighed. The decision maker will then have some firsthand knowledge of matters relating to the problem rather than having to rely entirely on a summary, which never gives a complete understanding and often gives an inaccurate one. In time of emergency, the need for prompt action and the great number of decisions that must be made force decision making to lower levels. If individuals at these levels are accustomed to making such decisions in peace, their decisions in war are much more likely to be prompt and sound. If they are accustomed to passing their problems up to higher levels in peace, they will delay and temporize in war.

17. Compliance with outside advice should not be required of the individual carrying responsibility for an operation. In the logistic field we receive many inspections, surveys, and reviews of our operations by outside agencies or individual consultants who are seldom informed on military matters. Although logistics is probably the closest military function to commerce and industry, only the military who have had experience in war can visualize the effect that a proposed measure may have. Improvements come, then, not from outside experts dictating solutions, but from their working with the military to develop solutions that reflect their combined knowledge.

18. The most useful type of report is one called for to help in the solution of a specific problem. In initiating such a

report, explanations can be offered as to the use to which the data collected is to be put. Only after a report has been policed through several reporting periods—its provisions thoroughly clarified and the explanations from reporting agencies for indicated shortcomings considered—will it serve its intended purpose well.

19. Where a specific important problem is known in advance and a reporting system has been developed to provide the data to solve it, that reporting system should be kept alive in peace. Otherwise changes in tactics, techniques, nomenclature, or accounting practices will render the reporting system unusable at the beginning of a new war. If the reporting system is not to be kept operating in peace, it should be taught in the Army's schools to ensure that the use of the data is known and the necessary explanations of interpretations that make the data usable are understood.

20. Reports not of sufficient importance to be frequently called to the notice of the commander to whose headquarters they are submitted are seldom worth the trouble of preparation.

21. A project officer should be an expediter who is to report rather than a czar who is to direct. The very existence of a second czar discredits the authority of the first, and the existence of several can only result in confusion. An absolute priority that would justify a czar is rare. In my time only the Manhattan project was worthy of one.

22. Cost-effectiveness is a valid consideration for military purposes only if the cost factor includes consideration not only of dollars but also of lives lost, lives blighted by wounds, and the effects of a national defeat. For commercial transactions, the cost factor is properly measured in dollars because the basic purpose of commercial transactions, is to make a profit in dollars. For military transactions the cost factor must be modified because the basic purpose of military transactions is success in war. Accordingly, effectiveness, in addition to having a relationship to dollars, also bears a relationship to casualties, wounds, and the successful outcome of a war. Since the value

of lives, health, and victory is difficult to determine, it is usually desirable to use cost-effectiveness only in deciding which of several roughly equally effective systems should be acquired. Where the systems are not equally effective, it is better to provide the best system or item that can be developed at its lowest reasonable cost.

23. Maintenance of material is so heavy a logistic burden that continuous effort should be directed toward reducing it. Maintenance requirements can be reduced by the following measures, among others: by a greater degree of standardization, by contracting for a higher degree of durability and maintainability, by making more parts throwaway instead of repairable, by improving the quality of rebuild, by transporting unserviceable material out of a theater of operations for repair, by prompt and adequate supply of repair parts, by care in operation, and by command supervision of preventive maintenance. All should be exploited.

24. The rate of consumption of repair parts for any new important end item should be determined as soon as the item is put into service and its accurace should be improved with experience. Repair parts consumption for any new item must be expected to vary widely from forecasts, regardless of how carefully made. Consumption will vary considerably with usage, age, and environment. Careful study of repair parts consumption through the useful life of the first models of any important new item is therefore necessary to establish requirements.

25. The impetus of supply is also from the front. The old axiom that "the impetus of supply is from the rear" is applicable to routine supply operations, but if supplies fail to arrive or if some special or unusual supplies are required, then a supply officer of the unit needing the supplies should go to higher supply echelons. If the shortage is critical, then his commander should go to higher command echelons to get action going.

26. The supply of seldom-used repair parts should be handled by cannibalization. In war there will be plenty of end items damaged by enemy action to provide sources for canni-

balization. For training in peace, undamaged end items should be cannibalized when damaged items are not available.

27. The movement of heavy equipment by air in the opening of a theater of operations is usually an undesirable procedure. In establishing a new theater, the requirement is usually not only for speed but also for strength to maintain a preponderance of force in the objective area. Strength is built up most rapidly by moving only personnel and light equipment (weapons up to and including 106–mm recoilless rifles and 4.2-inch mortars and transport up to and including ¾-ton trucks) by air. Heavy engineer equipment needed to build or repair airbases and ports is an exception. Armor, artillery, most heavy engineer equipment, transport over ¾-ton, artillery ammunition are best moved by water. It is well to plan, then, that operations requiring full divisions must await the unloading of the first shipment by sea.

28. All possible measures that can be taken in advance to speed unloading of sea transport in a new theater of operations are critical. The most dangerous time is when ships are being unloaded, both because the entire force is not ready to fight and because ships in harbor make a target highly vulnerable to attack by air, missile, or submarine. Among the critical logistic measures are the provision of landing craft, amphibious vehicles, and cargo helicopters; the use of roll-on, roll-off ships or containerized ships; the containerization of cargo; training in the pre-stowage planning for ships; and the training of engineers in beach operations, port clearance, and reconstruction, and construction and repair of airfields. Foreign military aid to a country on whose behalf the United States might intervene is probably the most valuable logistic measure that can be taken in advance. Even for a very weak nation, military aid is far more than just a measure to enable it to prove its willingness to fight in its own defense before the United States intervenes. Military aid to a foreign nation may also help to provide some invaluable time for the United States to prepare its expeditionary force; give some assurance that a beachhead will be held protecting port facilities so that a landing operation will be unnecessary; ensure that some artillery

and armored units will support our lightly armed troops that may arrive by air before the first sea shipments arrive; provide some stockage of transport and ammunition that can be borrowed until our own transport and ammunition arrive in adequate quantity; and instill some capacity in the troops of the aided nation to operate their own line of communications using and maintaining our types of equipment to save us from having to provide the logistic troops to perform these tasks for them.

29. An operating production line that can make the best end item so far developed with the best production equipment is an invaluable military resource. It is unlikely that the United States will start a war in the foreseeable future. We must therefore expect to react to the initiative of other nations. With an Army most like being a mobilization base, rather than a force in being, this takes time. No administration has been willing to finance the establishment of reserves of materiel large enough to last until the rate of production can catch up with the rate of consumption once mobilization occurs. We must therefore rely primarily on production initiated after the emergency appears imminent. The production of materiel takes more time than the organization and training of troops. Anything that speeds up the provision of materiel therefore speeds up the time when our strength can be brought to bear. To know in advance what we want to produce and how to produce it is, under these circumstances, far more valuable than stocks of obsolescent materiel.

30. Time, the essential measure of logistic readiness (as well as for everything else), should be bought by moving up the beginning of logistic action. Napoleon said, "you can ask me for anything you like, except time," and Benjamin Franklin said, "he that reseth late must trot all day." Although Franklin omits that he may not arrive at all, he comes closest to making the point that is not given sufficient recognition, namely that we can provide time by starting early. Thinking a possible operation all the way through and making a tentative plan, however inaccurate, may reveal areas where information is now firm enough to initiate some action; may point up possible

problem areas where study of alternatives or even initiation of research now will save time later; and may help in the recognition of the degree of accuracy of information, the degree of actual availability of resources, and the degree of authority to proceed. History shows what a nation threatened with war did in advance to enhance its preparedness; what it might have done is conjecture. But we in the military need to orient our minds more to identifying what proved to be critical early in each war and what the Army might have done, regardless of current political obstacles, to improve the situation.

31. A logistician must not only have integrity but also complete freedom from any suspicion of a conflict of interest. Throughout the military services, integrity is essential to operational effectiveness. It is not enough, however, for a logistician to be honest. It is necessary that he so conduct himself as to ensure that there is no basis even for suspicion that anything other than the best interests of the military service and of the Nation has been allowed to influence any official transaction.

Appendix

APPENDIX

Background of the Author in Logistics

Pre-1941   No logistics other than the little involved in peacetime service with troops or learned in the Army schools, except for one year, 1931–1932, at Purdue University, studying automotive engineering.

1941   Action officer in G–4 Division, War Department General Staff.

1942–1944   Planning Division, Headquarters, Army Service Forces. Director, 1943–1944. Logistic support from continental United States for overseas operations in World War II.

1944–1945   G–4, Mediterranean Theater of Operations. Logistic support in an active theater of operations.

1945–1948   G–4, European Theater. Clearing up logistic aftermath of World War II. Designing the line of communication for our forces in Europe in the event of another war.

1948–1949   Deputy Chief of Staff, European Command, and Chief of Staff, U.S. Army, Europe.

1949–1951   Deputy Undersecretary of the Army and Chief, Office of Occupied Areas. Support from U.S., primarily logistic, for the occupations of Germany, Austria, and Japan.

Chronology

1951–1952   Deputy G–4 for Plans and Programs, Department of the Army. Logistic support from the continental United States for United Nations forces in the Korean War.

1953   Commanding General, 24th Division, Korea.

1954   Commanding General, IX Corps Group, Korea. Provision of corps troops for the organization of the Republic of Korea V Corps. Resettlement of refugees into war-devastated areas.

1954–1955   Chief of Staff, Far East Command.

1955–1959   Deputy Chief of Staff for Logistics, Department of the Army.

1959–1961   Commanding General, United Nations Command, U.S. Forces, Korea, and Eighth U.S. Army. Establishment of a separate Republic of Korea line of communications. Enhancement of the ability of the Republic of Korea to earn foreign exchange.

☆ U.S. GOVERNMENT PRINTING OFFICE : 1990 – 259-883 QL 3